# TALKING THE TALK?
## How many ways …
## … can something be shared?

# TALKING THE TALK?
## How many ways …
## … can something be shared?

*Trade Languages …*
*And Trading Languages …*
*Through Performance Skill Training!*

## Melvin Pohlkotte

**To order additional copies of this book, contact:**
Xlibris
844-714-8691
www.Xlibris.com
Orders@Xlibris.com
812618

# CONTENTS

# PREFACE

In the 1970s, the 1980s, and the 1990s, we were blessed with the humor, the antics, and the unusual logic of a great comedian, Gary Wayne Coleman. One of the more famous questions that Gary shared frequently was, "What you talkin' about?"

Mr. Coleman died in 2010, but his persona, his theatrics, and more vividly his question lives on. I open with this tribute, because for so much of my personal career, that question has confronted me as I set forth my position on skill training.

Although I have enjoyed the benefits of education, I learned within the first few years after graduating from an engineering school, just what I had not learned. I did learn about manufacturing, and the role of various segments of the manufacturing organization. I did learn about the function of maintenance and the different accountabilities of operations, and administration.

I did learn about the role of engineers, marketers, and executives. I also did learn that the role of maintenance departments was to keep production processes and equipment in operation, *assuring maximum up-time*. What I did not learn in school, or even gain a hint of in the many textbooks, the many complexities involved in that simple expression, *assuring maximum up-time*.

I had completed 4-years in a co-operative program so I had become quite familiar with the workings of the operations department of a major automotive manufacturer. Although I knew about the presence of the maintenance department, and the work order process for requesting maintenance attention to a concern, I had no clue about a maintenance organizations procedures or problems.

My ignorance was short lived though, because my co-operative program required my completion of a one-year research and thesis project to fulfill my graduation requirements. I had been sponsored by the engine plant, but there was no agreement on a thesis title. So, I applied for a position and a research opportunity in the foundry maintenance department.

My first lesson in remediating my ignorance was enunciated immediately on my first day as a supervisor in the foundry maintenance department. Whereas I had spent 4 years in an environment that involved one instructor, and 30 students, I became painfully aware of the fact that now, I was in a class of one student (me), and 32 instructors (my employees).

To the man, these 32 employees would ridicule my accent or enunciation of the words as I attempted to read the work orders that had been submitted by production overnight. I did not have a difficulty with the words that had been used, my difficulty was with the word phrases that were used, and the way they described the objects that needed to be worked on.

Bear in mind, my prior visits to the foundry were to the office, to be interviewed. I had appeared for those interviews in by best clothes and neatly polished shoes, so the management did not feel it would be appropriate to take me on a tour. On my first day, I was in mechanics laundry blues, and wore company prescribed steel toed safety boots.

Tours were no problem, every employee wanted to show me their interpretation of the worst place in the foundry. Every one of them knew of hot places, dirty places, noisy places, etc. It only took two of these guided tours to persuade me that I was on trial.

Among the 32 employees, I had two blacks, one native American Indian, and 4 Hispanics. So along with the personally guided tours, there was quite a variation in language arts, and particularly with dialect, and colloquial expressions.

That was the first few days. Within the first two weeks, we had a severe breakdown on a piece of equipment in the core room. While considering what to do and who to seek out to make mechanical repairs, the maintenance superintendent suggested that I (being an engineer) visit the trade school on the property, and see if they could undertake the task.

I had been with the company for a bit more than 4-years at that time, but I was not aware that the company ran a trade school. To be honest, I was shocked that the company would run a school that was competing with the corporate college that I had attended.

I was more shocked the moment I set foot inside the trade school building. I am not certain that I expected to see drafting tables and file cabinets, but I absolutely did not expect to see one of the best equipped machine shops of that day. Nor did I expect to see nearly every machine being operated at one time.

I was extremely pleased when the trade school director immediately accepted the offer to machine a replacement part for our machine, and to undertake the restoration of the machine as part of their training. Within weeks, the machine was up and running, and I had become acquainted with the management of the trade school, and several of their students. Through those associations, I began to learn about job roles that had never been discussed or even hinted at in my years at school.

But those brief exposures to the trade school people were only my interview so to speak. As the trade school administration learned more about me and my education, they asked the foundry manager if I could be available to assist in developing curriculum and performance objectives for their students. That was agreeable and during the negotiation we determined that the trade school would like to build or repurpose foundry equipment as a means of providing *real world experience*.

Suddenly, the question, "What are you talking about, *real world experience*?" Do we not live in the real world, and what type of experience do we have if experience is not real world? Another shock, cracking my armor of ignorance. Although my father was an engineer, I grew up in a rural community in a town with a census count of 180 people. We did have a small grocery/post office combination. Oh, and a telephone exchange managed by humans with plug in jacks to connect with callers, and those they were calling. No street lights, one major intersection, and the simple life.

I was always curious about things that my father was doing on the job, and often times I would discuss what I was studying in science, or math with him. He would confirm what I was being told, but shared that I should not be surprised if I found out later that things do not behave quite that way in the real world.

Within a few short weeks of working with the trade school, my ignorance was totally exposed. Not only was the environment different from the schools I was acquainted with, so were the people. In math class in school many of our class had difficulty solving problems to the nearest hundredth. In the trade school activities, some of their apprentices were making measurements to the ten-thousandth of an inch.

In school it was not uncommon to witness cheating. In the trade school, no one worked closely enough to each other to cheat, and by intent the people closest together were working of different types of projects.

In school we talked textbook language (today that is typically identified as politically correct). In the trades program they talked shop talk, they did shop math, and instead of reviewing lecture notes or text books, they referenced trade related glossaries, job aides, or customer blue prints and specifications.

Seriously, day and night in the foundry, or the trade school, I was learning many things that I honestly thought I had learned in college. What I truly learned was the diametral difference between Academic Grading (norms, scales and curves), and performance evaluation against national standards (evidence based, and historically validated). Today that equates to the ***difference between education and training***.

# GRADUATE STUDIES

Looking back on that early start, I thank God that I had the privilege of a co-operative education, that I *earned while I learned*, and *did not incur educational debt*. In my formal education (COLLEGE), I studied for and earned a Bachelors in Industrial Engineering. But about my second year I came to the realization that the role of an Industrial Engineer in that era was to use a stopwatch, and a clipboard, and the observation of the activities of work, to establish performance standards, and establish the cost of manufacturing.

I was not excited about that discovery so I did some research into alternatives and began enrolling in Mechanical Engineering courses as overloads or electives. That took me into the fields of mechanization, mathematical analysis, and motion management. At my request, my job assignments were directed more to automation and continuous process improvement.

I also am so grateful for my enrollment in the school of hard knocks (the school colors were black and blue); of course, here I am talking about the foundry. In college, I learned the theory, and got to practice with a lot of simulators and toys. In the foundry, with a great crowd of witnesses and 32 instructors, I learned about

- real *terms*, ... words and phrases that were spoken to discuss a matter; differing words and phrases that were used to develop work orders and a plan of action; still different words and phrases used by suppliers to identify and source parts or replacement items. Frequently action words or phrases, unlike the descriptive words of the text books. The greatest sources for these action words were reference materials like the millwright's handbook or the Howard Sam's library for do it yourself clients (long before DIY became a marketing slogan).
- real *tasks*, ... complex work routines that frequently involved service literature research, part modification, cleaning, refinishing, and the use of a host of differing tools.

- real tools, ... real in this context meant appropriate. I had 32 maintenance employees spread over 12 trade specialties. No two toolboxes were equipped the same, and each owner took extreme pride in his favorites (usually wrapped in felt, or specialty tool holders. In that era, there was no such thing as tool checks and a tool crib for common hand and portable power tools or test instruments.

- real _techniques_, ... the way tools are used and the means of accomplishment. One of the most obvious variations in technique relate to the torqueing of bolts or nuts with a dial equipped torque wrench. Everything works well for someone that is right-handed. But the dial is not set up for a left-hander; so, a left hander has to work out his own technique.

- real _trends_, ... nothing is stable in the manufacturing automation arena. The only thing that is truly constant, is change. Quite often when you attempted to purchase a direct replacement, you found that the design had been changed, the features and benefits greatly improved, and the price lowered. These are all great news but the new part is not a bolt on interchange, so the entire assembly has to be reworked. Trends dictate future training needs as well as immediate inventory adjustments. You may not want to change, but an equipment manufacturer sends in a new piece of equipment and it is fitted with all new design product. You either change, or you risk paying a premium price for specifying the obsolete module products you are using.

- real needs for _tutoring_, ... this was one of the most difficult. As a graduate engineer with so called educational expertise in Industrial and Mechanical engineering concepts found myself on a daily basis asking for explanations as to practices, procedures and products. I did not have a help-desk, or ask an expert option, I asked the employees that worked for me. You can rest assured, I paid close attention, I did not want to go back and ask a second or third time.

- real necessity of advanced _training_, ... sometimes, you find yourself divorced from the think tank and you need upgrade training, or cross training for new skill demands. At that point you must pursue and acquire advanced training from a resource that has pursued the skillset you need and perfected it to mastery level.

- real personal performance *tactics*, this is the difference between the novice and the veteran. A novice will fish around for a means of accomplishment, a veteran will apply prior experience and provide information.
  - I recall hiring a new millwright with significant hydraulic experience. We had a trouble call on a molding machine (the system on the back of the machine was making an unusual noise).
  - I took the new hire to the site, and he looked, felt, and then said I will be right back. He returned from the shop with a mechanics stethoscope, and began to listen to various parts of the system.
  - A veteran mechanic with many years in the foundry saw this activity, walked over to the reservoir and took a wood pencil out of his pocket. Without saying a word put the pencil on the discharge pipe of the pump, put his ear down to the other end of the pencil, stood up, and said "the pump is cavitating".
  - We disassembled the suction line, drew the suction strainer out of the reservoir and found two problems;
    - The suction strainer was covered with debris to the point that it had very little open area,
    - This suction strainer was not equipped with a bypass valve that would have let dirty oil into the pump, without causing cavitation. We replaced the incorrect strainer and all was well.
  - *Tactic?* Perfected intuition and repetitive practice (the product of experience).
    - The one man had $100.00 invested in a stethoscope with its optional probes and a nice

carrying case, the other a $.50 pencil. Either could work, so neither was wrong, just different.

So, what did I gain from my enrollment in the school of hard knocks?

- Experience, and that is what you get when you expect something else.
- *Introduction to the 8 T's that I have since adopted as my training methodology. Every new concept has to progress through the 8 T's in succession. Terms, tasks, tools, techniques, trends, tutors, (more) training, and tactics.*
- I learned that what I had gained through 4-years of college was the theoretical *of the first three T's (descriptive).* The last five came through activity, behavior, initiative, etc.
- In essence I learned not only how to, "Talk the Talk" but I mentored, and tutored until I learned to "Walk the Walk".

More recently, while discussing communication with a class of students, I shared this expression of the talk and the walk. One young man raised his hand and said, but it is also important to know the impact of each, then he shared;

*Your talk talks,*
*And your walk talks;*
*But your walk talks*
*Louder than your talk talks!*

So true. And it gives credibility to the expression, "Out of the mouths of babes". I was the instructor, but in that moment a student shared a life changing view.

With that I will move into the meat of the matter, *Talking the Talk* so that we can *Walk the Walk.*

# CHAPTER 1

# TERMINOLOGY – WORDS BY WHICH WE COMMUNICATE

A seemingly unsolved question asks, "What comes first, the chicken, or the egg?"

The answer to that question is of little relevance to this discussion, but when we had a farm, we got chickens from the co-op so we could have eggs. There are however many chicken producers in our area that buy eggs so they can have chickens to sell. It seems to be a question of personal perspective.

The focus of this book however does parallel that question in the fact that when it comes to training, what comes first, the math or the reading. I will share that for the purpose of this discussion the reading, words, vocabulary, glossary, dictionary etc. comes first. Without words, we cannot even discuss numbers, and *without words and numbers* there can be no math.

Words are descriptive, and may be used to discuss numeric relationships such as six-pack, dozen, or gross. Numbers on the other hand are quantitative, they assign value, or establish a count. 6 in a six-pack, 12 in a dozen, 144 in a gross, and so on. But then there is also a half-dozen, or 12 dozen, and so many other ways to confuse.

Communication in the workplace is critical. The more specific the focus the more critical. A tour guide leading a group of visitors through a facility talks one way. The safety manager addressing a class of new hires will talk in a totally different manner.

Our focus in this book is on the skills necessary to operate and maintain process reliability in automated manufacturing. This includes everything from 1940's style linear transfer and electro-mechanical equipment controlled by push buttons and selector switches. But it also includes the most sophisticated robotics controlled by cell phones from many miles away. And not to be ignored, it includes every form of technical improvement introduced in between.

It is true that math is foundational, and a fundamental need, but we must expand our understanding of shop and *trade math* to include many forms of numeric expression. In any given meeting we may be using numbers to discuss production rate in pieces per hour. The next topic may delve into the quality of those pieces and be discussing the need to maintain tolerances of a specific diameter to one one-thousandth of an inch.

Then in that same meeting the discussion can turn to the price per piece, per dozen, per hundred, or per batch. Along with price, we might engage in discounts. In all of these discussions, however there is only one that would be classified as *trade math*; the tolerance specified as *one one-thousandth of an inch.*

In the following chapters we are going to investigate and discuss trade distinctions that must be mastered to effectively maintain reliability in the performance of modern manufacturing process machinery and equipment. We will not only deal with the fundamentals of a variety of trades, but with the gray area that resides between the trades as we integrate the disciplines.

By intent we will focus on mechanics, mechanisms, and motion management first since that is typically the means of manufacturing. Then we will discuss electrical and the nuances of electro-mechanical that are more recently giving way to electronics and instrumentation thanks to the space program. This electro-mechanical has been the mainstay of industrial control for more than one hundred twenty years.

We will also turn our attention to electronics, and consider the impact that technical advances have totally changed our capabilities, and ultimately to instrumentation and the impact advances in that field have totally changed our lives and lifestyles.

Although fluid power is considered part of mechanical, we will address it separately because the characteristics and attributes are so vastly different, and the availability of adequate training is becoming more and more difficult to source.

We will speak to all of these matters as an application technician, from the aspect of the tool's carrier, the mechanics laundry rented uniform wearer, the grunt. Not as the theorist. We will not speak from the textbook perspective. We will speak from the standpoint of more than 68 years of experience.

Our purpose is not to share information in the form of rules (legalistic) and regulations or as procedural guidance for personal performance. Our purpose is to share tactical experience and elderly advice, knowing how many mistakes we have made personally or witnessed. All such mistakes might have been avoided if we had only known then (when they happened) what we know now, as we write.

In an earlier series of books (eight in number) we compared the systems of the human body (things we either know about, or know someone that knows about them) to the operating systems that we call automation.

We began with the ground on which we reside and upon which we build. The terra-firma that stops us when we fall. The source of the gravity that often makes us fall. That ground becomes the anchor for our footing and the stability for our foundation.

We compared our feet on the ground to the footing of a build, and the source of stability in our foundations. Our feet then become the very foundation on which we rely for our mobility and productive activity. The foundation of the machine is comparable although totally different in size, and shape; and totally incapable of movement with regard to the ground in which it resides.

There is one similarity, when a foundation for a building is complete, we usually backfill the dirt around it so it is obscured from vision to the outside. We commonly cover our feet with shoes. In general, with regard to industrial automation, that is where the similarity ends.

Do not get me wrong, there are broad categories of mining, material handling, agricultural, aviation and aeronautic automation that involves mobility, but we are focusing on fixed, industrial, manufacturing automation. We will submit that mobilized automation incorporates the same concepts as fixed once you get above the foundation.

The foundation is truly the starting point. Everything we call automation is built on the foundation that we have structured. So, the next concern is fastening. How do we attach to the foundation?

In general, there are two broad categories into which fastening falls; within those categories industrial fastening is expanded to include;

- Permanent
    - Welding
    - Riveting
- Reusable
    - Threaded Fasteners
        - Bolts and nuts
        - Machine screws
    - Pins and cotter pins or spring clips

The sub-category *Threaded Fasteners* offers the greatest concern in industry by virtue of similarities in appearance, but critical differences in attributes and properties. This book would grow to be unmanageable if we attempted to put all of the information that we intend to share in print, so we will inform you on many of the essential data bits by way of Internet Search URL's.

The first of these submissions, relating to Threaded Fasteners and prompted by Gary Coleman's question "What are you talkin' about?

A GLOSSARY OF THREADED FASTENERS

**https://www.fastenerblackbook.com/index. php?app=fastenerterms**

This is an excellent resource, but it is one of many, and all of them that focus on terminology will seem to be repetitive. That is because terminology discusses the properties of the fastener (what it is). These are typical textbook words but enhanced by a bit of context. The pictorial enhancements in this glossary clarify some of the more difficult concepts, but with all they do they only answer the questions regarding what they are.

The questions that are so crucial to fastening in automation include;

- What does it do?
- What are the properties that qualify it?
  - Strength
  - Hardness

We could engage in numerous discussions as to the application of threaded fasteners used to mount equipment. There are two physical properties that distinguish the quality of a fastener and its ability to maintain a solid joint, grade, and thread form.

Then there are application concerns. As critical as these concepts are, they are among the first technical concepts that a child is introduced to. The topic, "*Simple Machines*". One extremely well-done web offer is available at the following URL.

**https://www.kyrene.org/cms/lib2/AZ01001083/Centricity/ Domain/968/The%206%20Simple%20Machines.pdf**

Many, however, prefer a much more academic approach to skill mastery another good site is available.

**https://sedl.org/scimath/pasopartners/pdfs/machines.pdf**

We will not attempt to add to the information available through the study of these two websites, but we will attempt to point out what is not shared in these sites.

Threaded fasteners are used for numerous applications that are involved in applications other than fastening. There are literally hundreds of applications that vary from leveling (consider the mounting feet of the domestic washer and dryer, or the refrigerator). In the industrial environs, there are many applications where threaded fasteners are used to establish a stopping point for something in motion.

Then there are fasteners used to adjust the alignment of belt pulleys, or coupled with a spring to adjust tension on a belt or chain take-up. A threaded fastener may be applied in controlling the flow of material through a valve or gate by controlling the size of the opening.

## ONE PHYSICAL CHARACTERISTIC THAT NEEDS EXPLANATON

The first of these involves the difference between course threads and fine threads. There is a URL that offers a good explanation and contrasts the differences.

https://resources.tannerbolt.com/articles/
coarse-vs-fine-thread-what-thread-type-do-i-need/

Then there is a site identified as Bolt Science that deals with frequently asked questions. This is a valuable demonstration of the value of experience. Experience only poses the questions, it takes research to find the answers, and it is impressive with people who find those answers are willing to share them. Study these questions.

https://resources.tannerbolt.com/articles/
coarse-vs-fine-thread-what-thread-type-do-i-need/

# THEN, ONE OF THE MORE PROBLEMATIC CONCERNS

The difference between English and Metric Fasteners. The following three URLs will provide a great understanding first of the history, then of size comparisons and then a delve into the performance properties of comparable sizes.

Be assured, with our global economy today, a maintenance mechanic must be bilingual when it comes to threaded fasteners and their selection and use.

http://knowhow.napaonline.com/metric-vs-standard-the-nuts-and-bolts-of-nuts-and-bolts/

https://www.engineeringtoolbox.com/screws-metric-inches-equivalents-convert-conversions-d_2056.html

https://www.albanycountyfasteners.com/Fastener-Reference-Tables-s/1157.htm

Threaded fasteners are but a tiny piece of the automation puzzle, but if they do not work as expected when applied, nothing else works as intended. As small a piece of the automation puzzle as fasteners are, they provide some of the most stressful concerns for purchasing and inventory control. Mix and match becomes a nightmare.

The worst of all concerns with regard to industrial grade fasteners is their local accessibility. The local hardware store and most big box stores do not stock industrial grade fasteners. This imposes significant stress of the off-shift personnel. Production management does not understand the words, "*It will have to wait till tomorrow*".

**EXERCISE** (For the fun of it)

A threaded fastener specified as follows …

*¼ inch, 2-inch-long, 28 thread per inch* socket head cap screw …

is used to move a stop for a tray in a bakery oven.

Question;

How far will the stop move away from the fixed barrier if the cap screw is turned 180 degrees clockwise.

Your answer _____

## COMPLEX MECHANISM

Spend some time with the following URL and gain an understanding of the meaning and impact of work by virtue of the mechanical advantage of each of the six simple machines, and then consider magnifying effect as these simple machines are paired up to do more with less and provide repetition through motion management.

### https://energyeducation.ca/encyclopedia/ Mechanical advantage

**Oops,** I let it slip, I let the word energy creep in without giving you access to a glossary that explains the words that are unique to the topic under consideration. Spend some time with this one, it is foundational.

### https://energyeducation.ca/encyclopedia/ Navigation:Index

Of course, these are both differing pages of the same website, but the keyword Mechanical Advantage would let us spend time on the site, and totally miss the Index.

Please let this be a lesson, and remember our discussion in the preface about starting with terminology first. We will honestly understand more of the discussion of mechanical advantage, if we cross reference the index.

We have implied that fasteners are typically utilized to mount and anchor equipment to a foundation, or to serve as an alignment device, an adjustable stop, or a positioning means.

Quite the contrary is true. There may be two to three times the number of threaded fasteners used in the actual automation than in the mounting. Some styles will be different, but as we progress through our discussions we will attempt to account for the many uses of threaded fasteners. Please refer to the Appendix, for additional resources on fasteners and fastener applications.

In case you haven't guessed, one of my goals in sharing this book is to encourage you to engage in meaningful web research for information that you need and will use.

It is important to know that beginning in the late 1980s and early 1990s major manufacturers were embroiled in mergers, acquisitions, and oftentimes hostile takeovers. It was economically impractical to attempt to merge the service literature of the many integrated trade names and product literature in a paper format with the advances in technology that were available.

The automotive and mobile equipment manufacturers took the lead and offered their service literature across the internet. If anyone is involved in maintenance of anything today (2020) they had better be web literate. Otherwise, they run the risk of inflicting damage to equipment and even posing a threat to the health and safety of personnel. Without up to date, validated service information, and evidence-based advice and counseling, we are winging it.

In the next chapter, we will address the major disciplines that are essential to the reliable management and maintenance of industrial manufacturing process reliability. In addition to identifying these disciplines, we will provide the basic foundational information for each discipline, beginning with a glossary of terms, and reliable resources for training, mentoring, coaching, and guidance.

There is no room in industry today for the self-declared *know it all*. A person like that is the source of an accident waiting to perform out of memory or on a suspicion, unwilling to seek the advice of reliable resources.

Before we move on to our next exploration, spend some time to investigate this URL and gain valuable insight into the many variations of the simple machines are adapted to accomplish variations in motion patterns, velocity, range of motion etc.

**https://www.tec-science.com/mechanical-power-transmission/basics/operating-principle/**

Enjoy and then join us for our next adventure (next chapter) as we discuss the dynamics of discipline diversity when integrated into automation. As we follow a path closely parallel to the series of books, "If it is going to be, it is up to me", we have discussed the feet, the footing, the foundation on which we build.

Once the footing is cured, it is time to begin to assemble the superstructure.

# CHAPTER 2

# BUILDING ON THE FOUNDATION

The sole purpose of automating processes is to effectualize the utilization of energy, and the management of repetitive quality in product and performance.

There may be no similarity in appearance or operation between automated processes. But the sole purpose of automation, from linear mechanization to the most intricate robotics, is to effectively and efficiently manage the use of power and other resources.

Remember, one of the last URLs that we shared was a glossary dealing with energy among many other things. It is one of the better glossaries available or the range of discussions we will be engaging in. Please use it often.

**https://energyeducation.ca/encyclopedia/ Navigation:Index**

Take the time to look up the following keywords so that the explanation that follows will be meaningful. You can make notes here, or start a logbook for these studies.

- Ohm's law
- Volt
- Amp

- Watthour
- Kilowatt hour
- Megawatt hour
- Energy
- Power

## TRANSMISSION AND TRANSFORMATION OF ENERGY

Since the purpose of automation deals with energy, this is the best place to start. Where the wires from the utility company come onto the industrial property. Sometimes they enter the building directly, at one voltage level. In the case of larger properties, they may enter a sub-station at a much higher voltage lever and be stepped down for the companies use.

The most expression of energy utilized by industry is ELECTRICITY. Although there are occasions that energy is purchased in the form of fuels, like gasoline for small engines, diesel fuel for larger engines, propane for fork trucks, and fuel oil for furnace operations. In these cases, we are talking about liquid or gas measure and we purchase these products by the gallon or by the barrel. The propane may be purchased by the cubic foot.

*ELECTRICITY* is purchased from the utility company in terms of *POWER* the timed rate of transmission of transformation of *ENERGY,* and is measured at the meter as *KILOWATT HOURS*, or *MEGAWATT HOURS*. If you look at an electric meter, there is commonly a disk or needle that turns at a specific rate and that rate is proportionate to the instantaneous use.

The meter reading that determines the charge levied by the utility company is totalized. That means that at any given instant, the needle or disk on the meter represent only a photograph of the process demands, the billed amount represents a video (an accumulation of usage over a prolonged period of time, expressed in hours).

The question that needs to be answered then is how does that energy no matter how measured, is "how does that energy usage translate to work?" And as was the case in the understanding of the purchase of energy, we must begin with the terms. The same glossary of terms is as good a resource as we will find, but we need to look up a new set of keywords.

Take the time to return to the energy URL and look up the following keywords

- Electrical power
- Electrical control
- Motor Control Panel
- horsepower
- Buss-bar
- Cable
- Ground
- Wire
- Electrical terminal
- Power
- Energy
- Work
- Resistance
- Efficiency
- Torque

All of these terms relate to the transmission of electrical energy from the point of entry to the many thousands of points of use. There is very little work accomplished in the transmission phase, because work requires effort, and distance.

The work accomplished in the transmission conductors is merely the work of overcoming resistance. The effort in electrical transmission is identified as a voltage level, and the flow rate is identified in terms of amperes. If there were no resistance to flow, the current would register at a higher level than what is normally experienced.

The difference between what could happen (theoretical) and what does happen (*real world*) measurements and recordings contributes to computations that we engage in to determine power losses, and system efficiency.

This raises another question, for those who remember their prior physics studies, and that relates to the loss between what is possible, and what actually happens. The textbook declares that energy can be neither created or destroyed: Energy can only be transformed. What the textbook rarely if ever acknowledges is that the losses always manifest themselves as thermal energy.

Think about your own experience with electrical appliances, particularly light bulbs. You buy light bulbs for illumination, either for personal comfort, or for effects. But I am certain that you have taken hold of a light bulb and burned your fingers. You do not normally buy light bulbs for heat, but you can. The difference is that heat lamps are not effective for lighting.

## IF NOTHING MOVES, THERE IS NO WORK DONE

Although there are losses in the transmission of electricity, we are incapable of measuring them unless electrical loads are operating, because there is typically no current flowing when the system is dormant.

With regard to electrical loads, we are going to limit our discussion to the applications of electricity that relate to automation and production processes. It has been shared that more than 85% of all industrial electricity is vested in the operation of electric motors of one form or another.

This may be a bit high for automotive or heavy equipment manufacturing due to the processes of welding and heating. If the use of industrial electricity is dedicated to the operation of motors, what do all those motors do?

One of the major categories of work that is involved in automated manufacturing is material handling (the movement of materials from one point to another). This is often done by means of conveyors. Conveyors are powered by belt, chain, or gear systems (open systems or enclosed gearboxes and transmissions). The choice of system is dependent on the load, speed, and the need for directional changes.

When any of these systems are idle, and the motor control equipment is deenergized (power off) there is no work being done. When any or all of the conveyors are operating, the energy at each motor shaft is being converted from electrical energy (supplied to the motor), to mechanical energy imparted to the belt, chain or gear system; and thermal energy (heat) radiated into the atmosphere by the motor and the conductor products delivering the energy to the motor.

When the systems operate, the energy delivered to a motor is discussed as kilowatts, but at the motor junction box we find ourselves discussing the capacity of the motor in terms of horsepower (the energy output terms at the motor shaft).

So, the electrical energy delivered to the motor junction box is measured and recorded as Watts (kilowatts) based on the voltage level and the current, proportional to the load demands of the system being driven. When we talk about the output shaft of the motor however, we must now talk about torque.

Torque is a product of force and distance. The force in our English system is expressed in pounds or other units of force, and distance is typically expressed in inches or feet (again in our English system).

## OUR FIRST INTRODUCTION TO MATHMETICAL CONVERSIONS

If we receive electrical energy expressed in watts, and we deliver that energy as ft-lbs of torque, and BTUs of heat, how can we reconcile these differences?

Quite often someone will approach this matter with the expression, "I think …".

Bluntly and being politically incorrect, it does not matter what anyone thinks. That is one of the problems in industrial reliability service supply today, too many do what think needs to be done.

There is a science behind all these concepts, and the resources that we have available to guide our activity is anchored in evidence-based and time-proven standards. In support of our discussion, we need to convert these differing forms of energy in terms of equivalency.

We can start with a formula that we will create on our own, because we have never seen such a formula in print.

## Electrical Energy = Mechanical Energy + Thermal Energy

No one will argue with that concept, or the expression of relationships and equalities. But expression merely uses words, general terms, textbook terms for that matter, they do not clarify anything.

So, to begin further analysis, let's add the typical units in parenthesis.

Electrical Energy (kilowatts)=Mechanical Energy
(lb-ft) +Thermal Energy (BTUs)

Unfortunately, there is nothing obvious in any of these that suggests commonality or ready conversion to terms of equality. Remember though, although our source of energy arrives at the motor as kilowatts, but the motor nameplate rates the motor in terms of horsepower. That is a term that gives us a means of conversion. If we look up horsepower equivalents in a good reference work, we will find the following;

| | |
|---|---|
| Electrical … | 1 horsepower = 746 watts |
| Hydraulic (and pneumatics) | 1 horsepower = 550 ft=lbs per second |
| | Or |
| | 1 horsepower = 33,000 ft-lbs per minute |
| Thermally | 1 horsepower = 42 BTUs per minute |

These units are called equivalents. To set forth the actual conversions we need to express our formula in terms of these equivalents as follows.

$$\frac{\text{EHP energy input -=(watts)}}{746 \text{ watts}} = \frac{\text{HHP energy output (ft-lbs)}}{33,000 \text{ ft-lbs/minute}} +$$

THP energy output (BTU)
42 BTU/minute

# WHAT ABOUT EFFICIENCY?

There are two efficiency considerations here.

ELECTRICAL EFFICIENCY =   ELECTRICAL ENERGY OUTPUT
ELECTRICAL ENERGY INPUT

HYDRAULIC EFFICIENCY =   HHP ENERGY OUTPUT
HHP ENERGY INPUT

MECHANICAL EFFICIENCY =   MHP ENERGY OUTPUT
MHP ENERGY INPUT

In round numbers, the electrical efficiency in industry typically ranges between 92 and 94%. Hydraulic efficiency in industrial applications typically ranges between 82% and 85%. For the record, pneumatic efficiency in industrial systems ranges between 45% and 60%. Mechanical system efficiencies in industry can run as high as 97% in rare cases but more typically in the 92-95% range.

There are people that say with hydraulic and pneumatic system efficiencies being this low why would the ever be used in favor of electrical? There are several hundred reasons for choosing belts, chains, gears, transmissions, hydraulics, pneumatics and many, many, hybrids: but that is the reason for this book. To promote pay-for-skill and structured training as an alternative to college.

The truth is that college will make a lot more sense, and be much more cost effective if the candidate has a trade certification first.

# WHAT ABOUT THE CATEGORIES OF WORK IN INDUSTRY?

One of the categories of work that is found in nearly every manufacturing environment in one form or another is material handling. This may mean moving automobile frames and chassis in one facility. It may mean transporting canned vegetables in another. Then there are bottled water and drinks, or all forms of candy.

The field of material handling not only involves linear conveyors, but it addresses elevators to accumulate product or to utilize multilevel physical facilities. Any of these applications of mechanism could conceivably be powered by belt drives, by chain drives, or by open or enclosed gear drives. All of these would be identified as MECHANICAL POWER TRANSMISSION.

In general, this type of material handling equipment involves long conveyors, fixed speed, single direction operation, and quite often continuous run, not short cycle start stop. Some of the advantages and disadvantages of these systems are as follows

- Belts are subject to slippage during starting, stopping; but they are capable of operating over a wide range of speed and loads within their design range, and typically require very little maintenance.
- Chains generally run at lower speeds than belts, but they can handle heavier loads, and are capable of directional change on a limited cycle basis. Chains are noisy, and operate best if lubricated. Lubrication however can become like flypaper in a dirty environment, and dirt and lubricant will accelerate wear of the chain and its companion sprockets.
- Gears are capable of managing heavy loads, running over a broad range of speeds, but need to be rigidly mounted, well aligned, and should be lubricated.
- The problem with all of these drive concepts is the need for a parallel or perpendicular relationship between the motor output and the drive output.

- In belt, chain and open gear drives the drive output most commonly needs to be parallel to the drive motor shaft.
- Compound gear drive packages (gear boxes and transmissions) are available for either parallel output, or perpendicular output.

By contrast, hydraulic, pneumatic, and direct electric drives can be configured to accommodate nearly any mounting (parallel, perpendicular or odd angle drive) requirements. Hydraulics and pneumatics have historically been the selections of choice where a wide range of speeds and frequent starts and stops were needed.

Material handling as a category of work generally implies moving materials and products from one location to another and is seldom involved with processing.

## ANOTHER CATEGORY OF WORK IS MACHINING AND FABRICATION

Since the early 1900s the machines used for metal cutting (machining) and fabrication (turning, milling, bending, pressing, extruding, drilling and abrasive machinery) have been powered by electric motors and driven through gear trains to enable speed and directional change.

As automation became a trend, significant numbers of accessories have been powered by pneumatic and hydraulic actuators. Many of these accessories are spoken of as material handling automation, as they are applied to load and unload or inspect a part being processed on a specific machine.

In terms of the energy transmission and transformation, the electric motor drives are the major contributors to energy usage. In evaluating the energy cost attributable to the operation of a specific machine, the drive motor is typically the only contributor considered. The accessory usage is generally calculated on the basis of the operation of a hydraulic power unit, or a pneumatic compressor.

# BUT WHAT ABOUT BUILDING ON THE FOUNDATION?

We have gone all the way around the barn to get to the point of this question. But there has been a purpose in the discussion so far. The motor, belts, chains, gears and fluid devices are applied to provide power and enable work.

There will be mounting brackets, conduit, hoses, and wiring mounted on the foundation and superstructure, but for the most part, the balance of the add- ons are what we call the automation, and they are weldments and fabrications that are considered to be engineered products.

All of the categoric products and practices we have eluded to thus far are based on the science of the performing disciplines. Mechanical drives, electric motors, hydraulics, pneumatics, and the practices of welding are all embedded in the sciences, and proven over more than a century of best practices and process improvements.

The superstructure, the framework and fabrications that we attach to the foundation, along with all of the articulating fabrications that we assemble to perform the tasks of work, are works of art. The imaginations of the designer, fabricated and applied to perform the many unique intricacies that parallel the abilities of the human body.

In one application, a clamp arm will swing down to hold a product in place. The next instant a slide will move to position the product in a machine, the machine will mill, drill, tap, or grind as programmed, and then the product will be returned from whence it came, the clamp will open, and the product will move on to its next adventure.

Weird or elegant in appearance, it does not matter, every attachment that moves, and every motion that each attachment moves through are applications of the six simple machines.

We discussed fasteners earlier, as they related to attaching the rigid framework to the foundation. Now, with the moving articulated members of the automation we will find many thousands of many different kinds of fasteners utilized to mount bearings, to lock bearings onto shafts, to position stops, and to mount switches and sensors.

The world today is vocalizing reluctance to accept the application of robots in industry. The truth is that automation from its earliest introduction has exhibit robotic characteristics. Modern robots merely look more humanoid, and instead of an arm motion being performed in one workstation and a wrist and hand motion being performed in another, the humanoid robot does it all in one station.

In essence, the robot has been applied to make better utilization of the floorspace, and has improve production cycle time by reducing the non-productive material transfer time between stations. Rather than a linear transfer line with workstations on either side of the line and material being bought to every point of use, the concept today is work-cells, robots positioned strategically to perform numerous tasks simultaneously.

The argument against robots is that they have **replaced humans**. The truth is they have **displaced humans**, forcing the skill levels up. Just as automation has required maintenance and servicing from its very onset, robots need to be serviced as well. The difference is each workstation in an automation may have involved the skills of one or two of the performance disciplines, the robot may involve as many as five or six disciplines.

While the electrician could service a workstation for a switch or sensor need, a millwright could service that same workstation at another time for a mechanical alignment need.

At another time, a hydraulic specialist could visit the workstation to attend to the reservoir, and adjust system pressures, or travel speed of hydraulic actuator.

But the service technician that responds to a call to service a robot needs to be able to deal with the total need (inter-disciplinary), or know who to call for concerns beyond their current level of knowledge.

## DISPLACEMENT, NOT REPLACEMENT

The greater the automation, and the more robotic we become, the higher the skill requirements in any environment,

The addition of the robot does not change the product that the robot is put in place to work on. Nor does it change the number of bits and pieces that are added to the product. What it does do is automate the material handling methods to bring the parts to the robot. There is no purpose in installing a robot and then telling the robot it will have to wait for a forklift driver to bring a tub of parts. Not only that, the sophisticated assembly or finishing robot is typically not fashioned to go to the tub of parts.

In the case of robotics, the parts have to be brought to the robot. What creates new equipment, new jobs, and higher-level skill requirements. Business and industrial stakeholders truthfully look to robotics to reduce or eliminate what has come to be known as grunt work for humans. This trend truly began when lost time and litigation became fashionable, for back injuries (repetitive moving of heavy product), hand-wrist and arm afflictions (repetitious abnormal movements).

We still have many of those grunt work positions in industry. Why are so many people reluctant to return to their jobs now that the government has made it more profitable to stay home? They know the government bailout is not going to continue, but they are not enamored by their current work.

Why are they wallowing in unsatisfactory work? Why don't they just go to school and prepare themselves for a better job? Simple, the shortest term from entry to diploma available today approaches two-years (in a post-secondary trade or technical school). Even with the current free tuition for two-year institutions offer, the cost will approximate $5000.00 for books, fees, travel expense, and that does not include lost wages if you attempt to go to school full time.

*Robots can only replace* if the *displaced person shirks* personal responsibility for continuous improvement of performance skill and technical knowledge. The most recent eight book series that are published on our website

www.aimtrain.net

entitled "If it is going to be, it is up to me" deals with this very dilemma. It is a current imperative that industry offer *pay-for-skill (earn while you learn)* opportunities to current employees (either unemployed or those still working), to upgrade the performance skills to the levels needed to service existing and new equipment.

It is also incumbent on the individual to pursue such opportunities with existing or potential employers. The book series and the online offers on this website offer a virtual toolkit for workplace readiness.

## DIRTY JOBS?

At the time of this writing, there is a TV program that runs on a weekly basis that is based on the theme, "It is a dirty job, but someone has to do it". On this program the narrator investigates all forms of jobs that are generally perceived to be dirty. Many jobs might fall into that category, but let us deal with one that pertains to automation, the topic of this book.

I am talking about an oiler, a person that is assigned the task of lubricating production equipment. This is only a dirty job because the wrong people are doing it, the wrong way.

It will never be anything but a dirty job until we change the name of the position, and deal with the incorrect perceptions in the minds of employers and career counselors.

The more attractive title would be LUBRICATION TECHNICIAN, and the position should be expanded to include technical decision making with regard to any oils grease, or artificial lubricants or additives used anywhere in a facility. In essence, this Lubrication Technician would be a TRIBOLOGIST. Look that one up in the dictionary.

Beyond lubricating equipment, this Lubrication Technician should be assigned the accountability for specifying and verifying correct hydraulic fluids for each machine or operation. All to often hydraulic fluids are stored in barrels, on a barrel rack with a spigot on the barrel so that oil could be drawn off into a can or pail.

Once a mechanic knows where a barrel of oil is, he gives no thought to the fact that this may not be the right oil for this application. Moreover, he does not get a vessel with a pour spout on it from a storage cabinet for clean vessels, he picks up the nearest pail he can find. It may have had speedy-dry (cat litter) in it the last time it was used, but it will still hold oil.

Next, he pulls the fill cap off and pours the oil in to that one and one-half inch hole. Folks, that is like raising a dump truck bed and trying to get all the gravel into a wheelbarrow. Not only does he miss the mark with a significant amount of oil, but he washes all of the debris that was around the fill tube into and off the sides of the reservoir, and he dumps what ever is loose in the pail into the reservoir.

The oil that missed the mark and ran down the sides of the reservoir becomes fly paper, and over a bit of time, contributes to the perception of oiling being a dirty job. It is only dirty because it was not done the right way, by the right person, with the right equipment, and with tender loving care.

It is like an old expression that I heard years ago in my *foundry education*. Many will share that *tools make the man*. That is not true, *tools make the living, the man just goes along for the ride.*

In the case of lubrication, this is a truth. Lubrication does not make the job a dirty job. The wrong people, doing the job the wrong way, and using the wrong materials make the job dirty.

Like a catch 22. This is an educational problem, but educators do not know that there is a problem. In their mind, any course on machinery operation and maintenance will give honorable mention to the need for lubrication, and in their minds that is sufficient.

The truth is, machinery does not need LUBRICATION, machinery needs lubricant REPLENISHMENT. That means the right lubricant, in the right amount, at the right time, administered the right way. Unfortunately, the mindset in industry today is, "That is not the way I was taught". And unfortunately, that does not change the facts one bit, obviously you were not taught.

Lubrication today is done on the basis of someone's opinion. In general, the old timer tells the new hire, put this gun on that fitting and pump till you see oil coming out of both sides of the bearing. WRONG, that procedure not only contributes to the perspective of a dirty job, that procedure literally destroys a bearing, because when you see lubricant coming out, you have destroyed the seals and the bearing is now exposed to the ambient and all its contamination.

You say we do not have airborne contamination in our plant. What about the humidity, airborne water? Water is not a lubricant. But how does water get into the bearing? Remember anything that does work is associated with movement, because of friction, anything that moves is inefficient, and generates heat. A bearing gets hot while the process is running, so it exhales the air from within.

When the process is shut down, the bearing begins to inhale, and draw new air from the atmosphere. And, the entire facility has heated up during the period of operation. Warm air holds more moisture than cool air, the capacity doubling for every 10 degree rise in temperature. So, there is water in the air that is now being breathed in.

As mentioned before, water is not a lubricant, but water is also not compatible with oils. Water is heavier than oil, so the water actually flushes the lubricant out of the bearing, and now iron exposed to water rusts and deteriorates.

There is truly so much more that could be shared, but this is intended to be a book, not a library, so we will offer more in variable forms as time goes on. Visit

www.aimtrain.net<br>and<br>www.mpsharecare.com

frequently to keep up with updates, new submissions, and breaking news as the TV networks advertise. Pay particular attention to the online offers, e-books, power points, and blog submissions.

SO WHAT?

I have been auditing a course perfected by a major university on EFFECTIVE WRITING. One of the suggestions is when you read something that offers a challenge or makes an argument, you should ask the question, "SO WHAT", and then read and reread the passage until you can answer the question.

I am not going to force you into that drudgery. I am going to answer that question here and now before we move to our next topic.

We have been addressing the topic, "BUILDING ON THE FOUNDATIONS", and have determined that anything built on those foundations that involves movement can correctly be identified as automation. That would even include guided vehicles (many used on transporting materials from a receiving or storage location to the many points of use).

The mechanization that we add to and anchor on the foundations is not the means of actually doing most of the work, they are the means of getting the tools and end-effectors (look that one up in the energy glossary that you have been working with) to the point of work, or meeting up with the materials or tools with which to do the work.

But why all the discussion of differing trades and particularly the lubrication technician?

Let's consider that question in depth, to answer the *SO WHAT* question.

Your organization has been working for months, digging holes, pouring concrete, laying reinforcing (re-rod) and pouring and finishing new floors. In many locations, there are either holes in the floor, or threaded rods sticking up out of the floor.

Now, the trucks arrive. Your organizations mechanics and millwrights along with the representative of the supplier begin to put fabrications in place and anchor them to the floor and foundation by means of those threaded bolts, or expandable anchors in the holes. Soon you have a stable frame.

Next, the millwrights and mechanics began to mount the super structure and add bearings, and pipe, hoses, and conduit all over the frame and structure. Soon, the hydraulic technician along with the mechanic start mounting cylinders, and valves, and connecting the hose and tubing.

The electrician works with the company representative to pull the wiring into junction boxes, and make the appropriate connections.

Then they mount sensors and run the cabling or conduit to the junction boxes, and they hook up the valves (pneumatic, hydraulic, water, or oil) as required.

When all of this installation work has been done, and the machinery (automation is ready) the electrician and instrumentation technician working with the factory representative program the machine, and a trial run is made.

There may be a misalignment mishap, and something gets broken, so now the welder is involved. Before long though, the machine is ready and after second trial run, turned over to production.

The machine is now running, and after months, sometimes years of planning and preparation, and the more recent involvement of the work of 6 or 7 trades, the machine is up and running. Great Work, folks, we could not have done this without you.

That is all well and good, and each of these trades' personnel have learned about the machine and its capabilities and nuances by working with the factory representative these many weeks or months. Beyond that, each of these trades' personnel will have many opportunities to apply their knowledge and demonstrate their skills hundreds of times over the next few years.

But my purpose in sharing and my response to your "SO WHAT?". It took the efforts of six or seven trades and the manufacturer's representative weeks or months to set this machine up and get it running. But the truth of the matter is simply this: One truly qualified TRIBOLOGIST (Lubrication Technician) will do more to keep this machinery running close to its *"as new performance"*, over the next 20 years than any other trade.

WHY?

Because lube replenishment done properly is always preventative. The response of all the trades is reactive. Trades always operate on the basis of, "If it is not broken, do not fix it".

This is not to discredit the trades; *the trades are what I am advocating for*. I am calling attention to this matter merely to declare that when lubrication becomes reactive it is already too late.

My singular purpose in these writings is to argue for the return to the days of American supremacy in productive quantity, productive quality, and personal pride in workmanship.

I am not advocating for a return to the old ways, rather a restoration of a process that accommodates enhanced knowledge, proven practices, and incorporates personal accountability for study and research.

We will leave this topic with two last URLs, a glossary of terms relating to footings and foundations, and a glossary of terms relating to fasteners.

https://en.wikipedia.org/wiki/Glossary_of_mechanical_engineering#top

https://www.fastenerblackbook.com/index.php?app=fastenerterms

# CHAPTER 3

## PUTTING THE MOTION
## IN AUTOMATION

If you have ever suffered an illness that sidelined you for a few weeks, you were probably shocked to find how much strength and loss of muscle coordination you had suffered during the period of recuperation.

As has been shared frequently in our earlier offers, motion requires muscle. The more the motion, the more the muscle. The other thing that is crucial to motion is the impact of inertia, and momentum.

Inertia resists movement. It takes more muscle to start a movement than it takes to keep it going. Momentum on the other hand, tends to keep motion going, so it takes more muscle to stop than it took to maintain motion.

We could go on with this type analysis of the need for muscle, but the fact of the matter is there are literally a hundred or more terms that impact our understanding of motion, and the means of managing motion.

Visit the following URL, and review the definitions, the implications, and the math associated with motion.

**https://biophysics.sbg.ac.at/glossary/physics.pdf**

In our previous discussions we have talked about the footings and foundations that are built from the ground up to support and sustain the forces of nature, and the impact of automation.

Then we talked about the framework which we attach to the foundation, to configure and maintain rigidity and alignment of the work to be done by the machinery and automation that we put in place to produce the products that we consume.

## WHAT THEN PROVIDES THE MUSCLE FOR THE MOTION OF AUTOMATION?

There are three primary means of providing muscle, **pneumatics** for high speed, low force applications, however, high-speed pneumatic actuation is a maintenance headache, because of the momentum of the actuator. A pneumatic actuator without cushioning or bumpers has the capacity to destroy itself.

Then for the brute force and modest speeds there is **hydraulics.** Although the hydraulic actuators are also capable of self-destruction, there are significantly fewer incidents of such because of the incompressible properties of the fluids, and the cushioning properties built in to industrial actuators. The third means of providing the actual muscle for movement is available through the application of **electrical** actuators.

Although we have referenced the glossary, we also feel it best to declare that the term actuator in any of these disciplines relates to linear motion products, and rotary motion products.

The primary producers of linear motion are typically called *cylinders, rams, dashpots, or bellows or diaphragms.* Similarly, the primary producers of rotatory motion are generally classified as *motors, or rotary actuators,*

# THERE IS MORE TO MOTION THAN THE ACTUATORS

As capable as all of these motion producers are, there are applications that have dictated hybrid integrations of interdisciplinary products to produce varying forms of motion patterns.

When we speak of linear motion, we are talking in general of only one form of motion. In all the applications that we know as automation, we encounter two forms of linear motion.

- Limited linear motion
- Continuous linear motion

And in all the applications that we know as automation, we also encounter two forms of rotary motion.

- Limited rotary motion
- Continuous rotary motion

One of the most astounding facts about motion management is that industrial actuators classified as *linear actuators* are not capable of producing continuous linear motion. And it is difficult to produce limited rotary motion, with precise control, with *motors*.

If you are not confused by now, consider this. The only way to produce *continuous linear motion* is to incorporate a motor and drive a wheel or gear on a surface or rack. To provide the most effective and manageable limited rotary motion is to incorporate a *linear actuator* with a *crank or a pinion* driven by a rack.

So, this gives us a brief introduction to the muscle for automation. The truth of the matter is that muscles need energy. No matter what form the muscle takes, the energy for this muscle originates with electricity. In most cases, an electric motor drives a compressor for pneumatics or a pump for hydraulics. When electric actuators are selected, the actuator is typically a threaded screw driven by an electric motor and a rod driven forward and back by the threaded screw.

# TO MEET THE NEEDS OF AUTOMATION, ACTUATOR MOUNTING OFFER DESIGN OPTIONS.

No matter what the means of powering the actuator, air, hydraulic, or electrical, there are many mounting options. Some promote linear alignment, to assure smooth efficient linear travel. Some however, provide mountings provide for articulation, and are used to provide limited rotary motion. Study the mounting patterns portrayed in the following chart.

## NFPA CYLINDER MOUNTING STYLES

| MH 0 No Mounting | MH 6 Intermediate Trunion Mount MT 4 | MH 14 End Lug Mounting MS 7 | MH 20 Cap Square Block ME 4 |
|---|---|---|---|
| MH 1 Foot Side Lug Mounting MS 2 | MH 7 Cap End Trunion Mount MT 2 | MH 16 Side Tapped Mounting MS 4 | MH 21 Cap Detachable Eye MP 4 |
| MH 2 Head Trunion Mounting MT 1 | MH 8 Head Tie Rod Extended MX 3 | MH 16 Side Tapped Mounting MS 4 | MH 22 Cap-Fixed Spherical Eye MP 5 |
| MH 3 Cap Fixed Clevis Mounting MP 1 | MH 11 Cap Tie Rod Extended MX 2 | MH 17 Head Square Flange Mounting MF 5 | MH 24 Cap Detachable Clevis MP 2 |

| | MH 12<br>Head and<br>Cap<br>Tie Rod<br>Extended<br>MX 1 | MH 18<br>Cap<br>Square Flange<br>Mount<br>MF 6 | MH 25<br>Cap<br>Fixed Eye<br>MP 3 |
|---|---|---|---|
| MH 5<br>Cap End<br>Flange Mount<br>Rectangular<br>MF 2 | MH 14<br>End Lug<br>Mounting<br>MS 7 | MH 19<br>Head<br>Square Block<br>ME 3 | MH 27<br>Cap<br>Detachable<br>Clevis<br>Duplicates MP |

Visit the following URL for a visual of these descriptions

https://www.google.com/search?sxsrf=A
LeKk02j3wE0BB-WxdicrZa4zhCjWICVRQ:
1601163931838&source=univ&tbm=isch&q=
**nfpa+hydraulic+cylinder+mounting**+styles&sa
=X&ved=2ahUKEwjh_oi7gIjsAhVGMt8KHZxC
ANMQjJkEegQIChAB&biw=1364&bih=665

With all of the mounting variations, it is important to realize that every one of the devices pictured in this chart, all work the same way.

In the event the actuator is either pneumatic or hydraulic, the construction involves a tube, end caps, and tie rods as housings then there is a piston, a rod, and a rod seal associated with the movement pattern.

By means of valving, pressurized fluid is introduced to one end of the actuator or the other. To move forward, the fluid is introduced in the blind end, pushing the piston and rod forward (the rod leaving the cylinder. When the fluid is introduced to the rod end, the piston and rod move back into the cylinder.

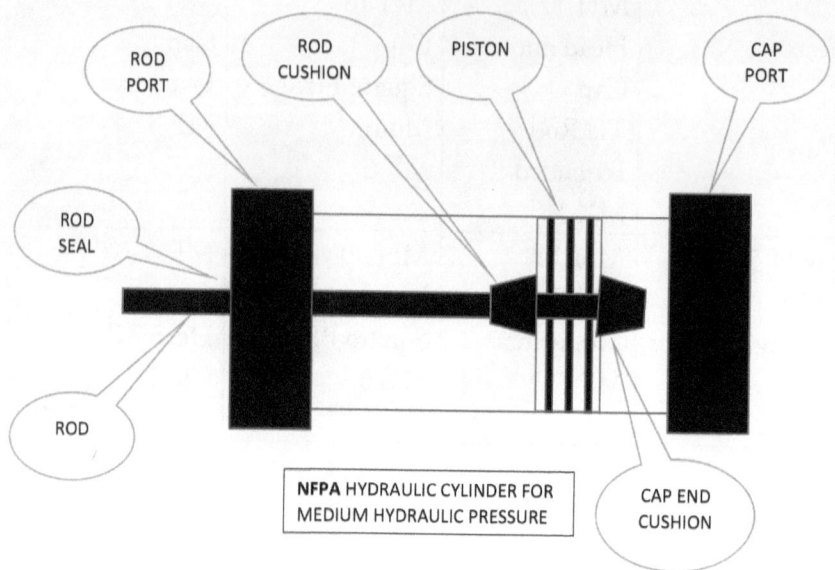

NFPA HYDRAULIC CYLINDER FOR MEDIUM HYDRAULIC PRESSURE

Visit the following URL for Actual Cutaway pictures.

https://www.google.com/search?hl=en&tbm=isch&sxsrf=
ALeKk00pTeTe2ahaej2WEV6WeWzxDeud
ng%3A1601165199033&source=hp&biw=11
50&bih=727&ei=jtdvX72rPKSe_Qawz4lI&q=nfpa
+hydraulic+medium+pressure+cylinder+cut
way&oq=**nfpa+hydraulic+medium+**
**pressure+cylinder+cutway**&gs_lcp=
CgNpbWcQDDoHCCMQ6gIQJzoECCMQJzoICAA
QsQMQgwE6AggAOgUIABCxAzo
GCAAQCBAeOgQIABAYUNxTWP3fA2CglQRo
FXAAeAGAAf8DiAHKLpIBDDQxLjcuMC40LjEuM
ZgBAKABAaoBC2d3cy13aXotaW1nsAEK&sclient
=img&ved=0ahUKEwi9qqaXhYjsAhUkT98K
HbBnAgkQ4dUDCAc#imgrc=1ZdCn72dp86MWM

The NFPA used in the identifying note stands for the NATIONAL FLUID POWER ASSOCIATION, and identifies a standard by which manufacturers build products that are considered equivalent, but the standard virtually assures dimensional interchangeability.

The electric cylinder appears similar, but the rod is hollow, a nut is welded or solidly affixed in the inboard end of the rod, a screw is fitted with a pulley at the rear for belt drive, a sprocket for chain drive or a gear for gear drive. When the motor is energized for advance the screw turns to extend the rod. When the motor is energized to return, the screw turns to retract the rod.

The positional accuracy and stiffness or load stability are more precise with the electric positioner than with any of the alternatives, but structural strength, and the ability to mount any significant weight on the rod is limited to designers' specifications. The rod of an electric actuator is by necessity hollow, and its strength is designed for push and pull applications.

The one drawback for an electric cylinder is the need for a servo or stepper motor and a companion controller to enable speed control or positional accuracy in motion management.

Whereas the pneumatic or hydraulic actuator and conventional directional valves would be classified as electro-mechanical. To provide variably controlled position or speed, the process would mandate instrumentation.

The electric actuator would be generally categorized as instrumental control by virtue of the power supply and the need for programmed instruction

One of many styles, and shapes of electric actuator. Notice the threaded screw in the rod. The brass ring around the rod is a bearing, and some models have a special construction to keep the rod from turning during the motion pattern. The term to identify this would be a *non-rotating rod*.

Typically, if the application of any linear actuator involves a load or process that attempts to rotate the actuator rod, it will be guided externally rather than rely on the cylinder to resist such. Think about it, the cylinder rod might resist rotation, but there would be no means of keeping the rod end connection from loosening.

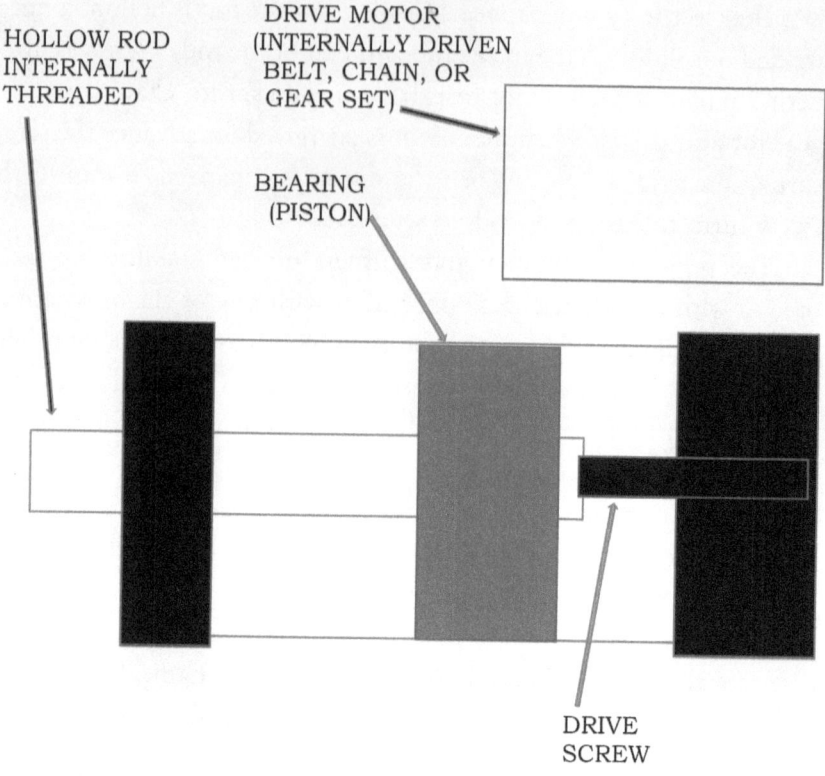

HOLLOW ROD
INTERNALLY
THREADED

DRIVE MOTOR
(INTERNALLY DRIVEN
BELT, CHAIN, OR
GEAR SET)

BEARING
(PISTON)

DRIVE
SCREW

Visit the following URL to see pictures of a wide variety of electric actuators. These are finding wide acceptance in light to medium load applications, and clean environments.

https://www.google.com/search?hl=en&tbm=isch&sxsrf=A
LeKk00pTeTe2ahaej2WEV6WeWzxDeudng%3A160
1165199033&source=hp&biw=1150&bih=727&
ei=jtdvX72rPKSe Qawz4lI&q=**electric+actuators
+images&oq=electric+actuators&gs** lcp=CgNpb
WcQARgEMgIIADICCAAyAggAMgIIADICCA
AyBggAEAUQHjIGCAAQBRAeMgYIABAFEB4yBgg
AEAUQHjIGCAAQBRAeOgcIIxDqAhAn
OgQIIxAnOgUIABCxAzoICAAQsQMQgwFQy0VYin
xgzqkBaAFwAHgAgAGNDYgBuBSSAQYx
Ny44LTGYAQCgAQGqAQtnd3Mtd2l6LWltZ7ABCg
&sclient=img#imgrc=PD5FZFcVGSU2aM

# ONCE MORE, THERE IS MUCH MORE TO MOTION THAN THE ACTUATOR.

It is not hard to see that all you would need to push or pull a load with the electric actuator once the mounting, guiding, and attachments were connected and the power supply was mounted, would be an electric supply, and a program.

The popularity of the electric cylinder has grown rapidly since legislation has mandated the need for public and service organizations offer handicap services relative to swimming pools, therapy pools, etc. the electric cylinder, an electric supply and a pendant for the patient or caretaker, and you are in business.

In the case of *pneumatics*, you need a **compressor**, a *receiver* (holding tank) ***piping*** from the compressor to the point of use, and then ***local air quality servicing equipment*** (filter regulator and lubricator unit) ***a directional valve, flow control valves***, and all of the ***piping, tubing, hoses and connectors*** to assemble the system. That may well be why the output actuator is often called or taught as the ***final element***.

In the case of pneumatics, the application technician merely sizes the various components to accomplish the work at hand. It is assumed that the company compressor and distribution piping is capable of accommodating the added load.

In general, the compressor and distribution piping of a facility are considered to be a central system.

Hydraulic systems however are considered to be point of use systems like the electric actuator systems. The difference however, and much of the reason for choice is revealed in the fact that the many hydraulic actuators on a given system are commonly fed from a stand-alone hydraulic power unit.

A power unit is commonly comprised of reservoir, a suction strainer, a pump, a relief valve and a return line filer; with appropriate temperature and pressure indicators/gauges.

Then, like the pneumatic system, piping is run from the pumping supply to each sub-circuit. A typical sub-system involves a directional valve, then flow controls and the actuator. Often times, a hydraulic sub-system requires special pressure controls, back pressure controls, cross-line relief control, or counter-balance control. These are the exception rather than the rule. When these optional adds are needed, they only impact the circuit in which they are assembled.

No matter how elaborates the hydraulic system may be, on any given machine, each actuator and the servicing valving constitutes a singular sub-system.

## HOW MUCH WORK CAN A LINEAR ACTUATOR DO?

In general, material handling and product transfer, clamping a light machining or assembly operations do not require extremely heavy loads. These categories of work would typically require less than 1000 pounds of force.

When the process involves metal stamping, bending, forming, extrusion, etc., the force required will be typically be expressed in tons of force. Sometimes in large panel stamping presses the tonnage is as high as 750 tons, but that is not often done with fluid power.

Just to illustrate the potential, without attempting to tie that potential to a specific task or job, let me share that about 85% of all of the fluid power cylinders used in industrial automation are in a range of sizes between ½ inch in diameter, to 8 inches in diameter. Since we are interested in the potential force, let us do the math, calculate the force that an 8-inch diameter actuator is capable of producing at varying levels of pressure.

It is not difficult to see that the ***Push force*** of an actuator (cylinder) is produced when the pressurized fluid is directed into the blind end of the cylinder. The piston is pushed away from the blind end of the cylinder. A ***Pull force*** is produced when the pressurized fluid is directed into the rod end of the cylinder.

For any given cylinder size, the ***push force*** will remain the same, but the ***pull force*** will differ from one cylinder to another, depending on the selection of different size cylinder rods.

Considering the calculations needed, the formula for the force produced by an actuator/cylinder is based on the formula.

$$Force = Pressure \ X \ Area$$
Where the units are
Force (pounds or tons)
Pressure (pounds per square inch)
Area (square inches)

## CONSIDERING AN 8 INCH DIAMETER CYLINDER

We need to begin by calculating the area of the cylinder. Area is determined from the formula.

$$Area = pi \ (\pi) \ X \ r^2$$
Where the units are
Area (square inches)
Pi = 3.1416 (a unitless constant)
$R^2$ is the radius (inches) squared
Then substituting
$$Area = 3.1416 \ X \ 4^2 \ inches$$
$$Area = 3.1416 \ X \ 16 \ inches^2$$
$$Area = 50.3 \ inches^2$$

Creating a table to display the forces potentially generated by this cylinder

| CYLINDER AREA | FLUID PRESSURE | FORCE GENERATED |
| --- | --- | --- |
| 50.3 square inches | 1000 psig | 50,300 pounds (**25.15 tons**) |
| 50.3 square inches | 2000 psig | 106,000 pounds (**53.30 tons**) |
| 50.3 square inches | 3000 psig | 150,900 pounds (**75.45 tons**) |
| 50.3 square inches | 4000 psig | 201,200 pounds (**100.60 tons**) |
| 50.3 square inches | 5000 psig | 251,500 pounds (**121.50 tons**) |

That is a lot of force from a piston approximately the size of a salad plate.

To put the difference in potentials in perspective, without having to wait another semester, or buy another book, let's compare an 8 inch bore pneumatic cylinder and the potential force generated by it at 100 pounds per square inch gauge pressure with a small diameter hydraulic cylinder and normal hydraulic pressure ranges.

| CYLINDER AREA | FLUID PRESSURE | FORCE GENERATED |
|---|---|---|
| 50.3 square inches | 115 psia | 5,784.5 pounds (**2.89 Ton**) |
| 1-inch hyd .7854 in² | 2515 psia | 1,975.3 pounds (**.99 Ton**) |
| 2-inch hyd 3.1416 in² | 2515 psia | 7901.1 pounds (**3.95 Ton**) |

Startling disclosure, a 2 inch bore hydraulic cylinder operating a 2515 psia (absolute) pressure is capable of producing 1.36 times the force of an 8 inch bore pneumatic cylinder operating at 100 psia (absolute) pressure.

WHY PSIA?

All fluid power calculations involving pressure should be based on absolute pressure, which is the sum of atmospheric pressure (14.7 psi at sea level commonly considered 15 psi or one atmosphere) the pressure of the air in the atmosphere from earth to outer space. And that added to the gauge pressure. The impact on hydraulic pressures in the thousands of pound range is so insignificant that it is often ignored in hydraulic computations.

SO FAR SO GOOD, BUT NOT ALL WORK CAN BE ACCOMPLISHED BY PUSHING.

| ROD DIAMETER | ANNULUS AREA | FORCE AT 1500 PSI (Pounds/Tons) | FORCE AT 3000 PSI (Pounds/Tons) |
|---|---|---|---|
| 8.00 INCH BORE | 50.30 IN$^2$ | 75,406/37.70 | 150,810/75.405 |
| 3.50 ROD | 40.65 IN$^2$ | 60,975/30.49 | 121,950/60.975 |
| 4.00 ROD | 37.70 IN$^2$ | 56,550/28.28 | 113,100/56.55 |
| 4.50 ROD | 34.37 IN$^2$ | 51,555/25.78 | 103,110/51.56 |
| 5.00 ROD | 30.64 IN$^2$ | 45,960/22.98 | 91,920/45.96 |
| 5.50 ROD | 26.51 IN$^2$ | 39,765/19.88 | 79,530/39.77 |

You can do your own math if you are interested in what percentage of the push potential each rod size offers in pull potential.

Why would anyone want to use the larger rod sizes? Column strength, structural rigidity, alignment, stability, and on and on. And, in most applications of automation, the actuator is applied in one direction of travel, and then return under low load or no load.

## SIDEBAR

I do hope that the depth of technical competence essential to the roles in reliable manufacturing process management and monitoring services is obvious.

And with all that we have shared here, in our earlier books, and on our website, I do hope you are beginning to get the true picture. One of the most promising career categories in America today is in skilled and semi-skilled positions in production process monitoring and management.

Moreover, recent reports declare that there are more than 7 million unfilled jobs in those categories. Now, we are starting to bring major industry back from offshore and reestablishing their presence on American soil. The need for skilled and semi-skilled workers to meet that need could easily double.

This will be my tenth book this year. Nine of them have been motivated by the needs of the vast population identified as un-employed, but more importantly, the even greater population that can only be identified as under-employed.

The truth is, possibly every living soul in America could be categorized as under-employed. We could all be more productive than we are, make a larger contribution than we are making, and influence more people for good than we are influencing. For years, biological and neurological scientists have declared that we humans rarely use more than 10% of the capacity of our brains.

It is truly a shame that such valuable resources are permitted to dwell in periods of idleness, or coasting. It would be such a valuable investment in one's personal life if they would passionately commit to lifelong learning. However, our energies need to be focused on pursuits that fulfill our creative purpose.

The great commission set forth for every living believer the obligation, not option, to make disciples and teach them life skills, and mentor them as they mature and go out and make their own disciples. There is no question that the lifestyle at the time was simpler, but human nature was the same.

There have always been the haves and the have nots, and it still exists because of that human nature. The haves keep and do not share; not out of obligation, but out of love. We grow up believing that if we do not keep what we get, we will ultimately suffer, but the truth of the matter is that those that willfully share prosper more.

We all have more to do, and I am committed to keep sharing the means of accomplishment without incurring significant debt, without struggling with institutional schedules, remote campuses, outdated textbooks, inexperienced instructors, and a system that openly declares grunt work is not our mission.

IT IS A DIRTY JOB BUT SOMEONE HAS TO DO IT.

It is true that industrial maintenance and automated process support activities can be dirty work as things are now. Due to the lack of basic technical skills, industry is attempting to rely on unskilled personnel to perform skilled tasks, and to exhibit the pride of ownership in the work area.

Political correctness would not permit them to post a sign in each work area saying, "Clean up your own mess, your mother does not work here". That would be met with the type of distain and claims of entitlement such as we heard from rioters and looters in Chicago recently, "There's nothing wrong with what we are doing, we are hungry, and we deserve the food". The truth is many of the participants were paid to do what they did by people from another geography. People who had no vested interest in Chicago, or Portland, or New York, or any other American city.

The Bible declares in the book of Thessalonians 3:10, "If a man does not work, he should not eat".

My purpose in sharing is to point those who are willing to a means of *earning while they learn, starting a bit lower* than the going rate for specific job categories, *and growing.*

Not only are we pointing a way by means of these books, but we are offering the resources by way of our website

www.aimtrain.net<br>and our blog site   www.mpsharecare.com

there are virtually unlimited resources on these sites.

BACK TO OUR STUDY.

We have been discussing the muscle that enables the motion in automation. So, let us consider a bit about the way the energy is conveyed from the source to the point of use, and the unique concerns that would promote one product over another.

## THE CATEGORIES OF ALL CONTROL

Before we can intelligently discuss and specify and source and select and supply products, we must come to an understanding of the function that each product will ultimately play in the overall of our plan and purpose.

Sometimes, words fail to convey a clear picture of the story we are attempting to convey. This is truly one of those times. So, in a manner that you will not likely see anywhere else, I am going to share the categories in two manners.

First, because as a reader, you recognize and understand the meaning of words, and the context in which they are embedded. Therefore, we will begin our exploration of control with words, appropriately placed, and then move to a pictorial representation that few are truly acquainted with.

## ENERGY CAN BE NEITHER CREATED NOR DESTROYED,

The energy that enables industrial manufacturing begins with transformation. Historically the electricity was produced by the transformation of fluid energy into electrical energy, taking advantage of waterfalls and gravity to turn turbines and generators. Then coal and oil-fired boilers produced steam, and that steam was used to drive turbines and generators, and produce electricity.

More recently, we have thermal, thermal, nuclear, solar, wind, and who knows what else that permit our transformation of each form of energy to electricity. With this history, and inventory of processes, we could be embroiled in endless search of how's and whys, when the truth is that all we need to know is simple.

The energy we need to run our manufacturing processes and domestic and creature comfort needs comes into our presence, our facilities, and our workplace on a set of 2, 3, or 4 wires. We now have all we need, to perform, the challenge to us is to manage it judiciously and efficiently so that we do not abuse our privilege.

Once in our facilities, our energy is utilized for facility systems, for powering our processes, and for controlling the management of those powered processes, our focus in this book is to provide enlightenment regarding industrial power and control, and demystify the unfounded perspectives associate with trades associated with the management of manufacturing processes.

## PROCESS CONTROL

No matter what the process or the means of applying product to provide control for all of the processes or manufacturing all of he concepts of control fall into four categories.

- Information
- Decision
- Output
- Final Elements

## INFORMATION

In this category the concerns are involved with *what is, what is wanted*, and *what happened*. What is, speaks to relates to status. What is wanted, speaks to the program, the sequence, the safety concerns, the quality concerns, and the unique intricacies and nuances of the locale and atmosphere. What happened, feeds back to the status providing a closed loop, step by step progression from beginning to end.

## COMMUNICATION IS MORE THAN WORDS

All our lives we have been acquainted with Braille as the language of the blind, and we feel the little dots on the elevator buttons in public buildings. And, from time to time, we witness deaf people speaking to each other in sign language. We have also been told for much of our lives that pictures speak louder than words.

One of the great shocks when someone enters into some training in some disciplines, and are introduced to symbolic languages and graphic communication styles.

The truth is that symbolic languages are the means of communicating the way things work (functionally). With regard to process control, and the categories of information, decision, output and final elements, our ability to share their relationship and meaning in words is severely lacking,

One of the most common symbolic languages of the trades is the Ladder Diagram, the foundational programming language common to the industrial electrical trades.

## BASIC LADDER LOGIC ILLUSTRATION

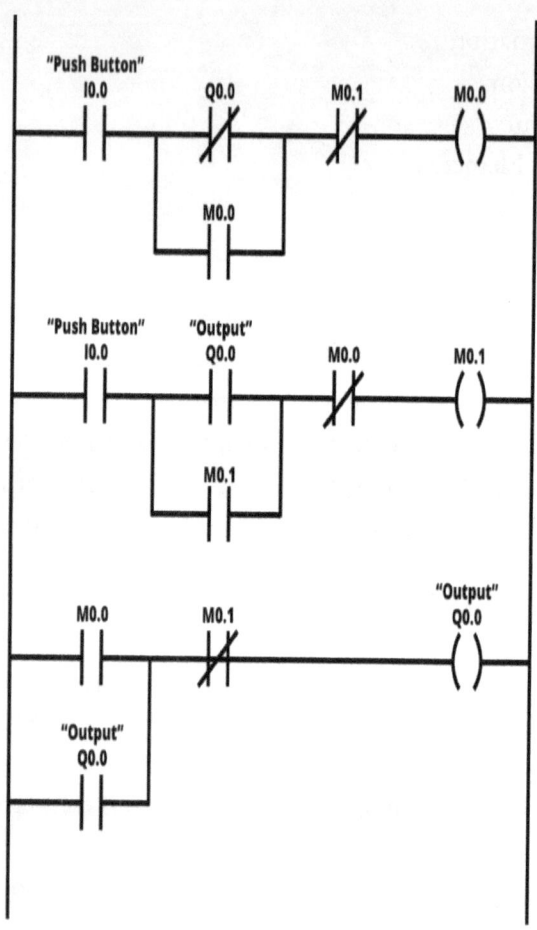

This ladder tells a story in three words; push-button, relays and output. But there are a lot of letters and numbers too. So, let's explain this by starting at the beginning. The existing picture illustrates the reason for the term ladder diagram. It looks like a ladder; two rails and several rungs.

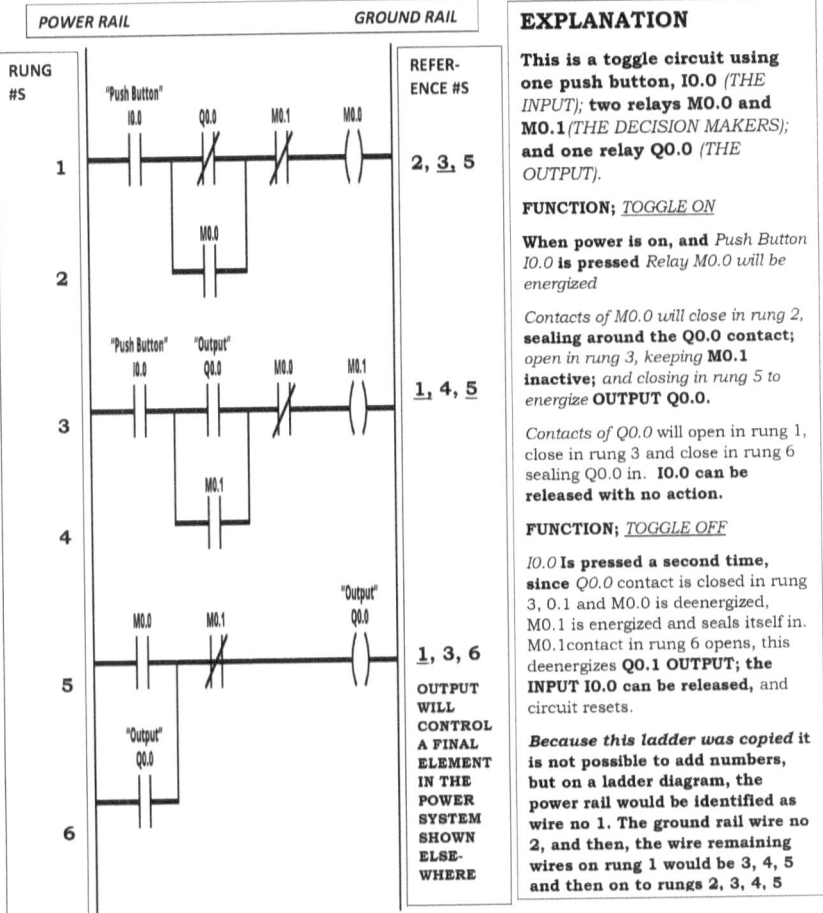

| POWER RAIL | GROUND RAIL | EXPLANATION |
|---|---|---|

**EXPLANATION**

This is a toggle circuit using one push button, I0.0 *(THE INPUT)*; two relays M0.0 and M0.1 *(THE DECISION MAKERS)*; and one relay Q0.0 *(THE OUTPUT)*.

FUNCTION; *TOGGLE ON*

When power is on, and *Push Button I0.0* is pressed *Relay M0.0 will be energized*

*Contacts of M0.0 will close in rung 2,* sealing around the Q0.0 contact; *open in rung 3, keeping* M0.1 inactive; *and closing in rung 5 to energize* OUTPUT Q0.0.

*Contacts of Q0.0 will open in rung 1, close in rung 3 and close in rung 6 sealing Q0.0 in.* I0.0 can be released with no action.

FUNCTION; *TOGGLE OFF*

*I0.0* Is pressed a second time, since *Q0.0* contact is closed in rung 3, 0.1 and M0.0 is deenergized, M0.1 is energized and seals itself in. M0.1contact in rung 6 opens, this deenergizes Q0.1 OUTPUT; the INPUT I0.0 can be released, and circuit resets.

*Because this ladder was copied it* is not possible to add numbers, but on a ladder diagram, the power rail would be identified as wire no 1. The ground rail wire no 2, and then, the wire remaining wires on rung 1 would be 3, 4, 5 and then on to rungs 2, 3, 4, 5

Reference #S for the rungs:
- Rung 1: 2, 3, 5
- Rung 3: 1, 4, 5
- Rung 5: 1, 3, 6 — OUTPUT WILL CONTROL A FINAL ELEMENT IN THE POWER SYSTEM SHOWN ELSEWHERE

**And that folks,** is about as fast a crash course on ladder diagramming as you will find anywhere; even faster than the Cliff Notes you may be acquainted with.

But because the relay on rung 5 is identified, it is obvious that there is more to the story ... somewhere.

We shared earlier that the electrical energy that was delivered to an industrial facility was utilized in three differing ways, facility systems, control, and power.

What we have just discussed and the most common means of discussing *control* is by means of the ladder diagram.

Facility systems are commonly associated with facility lighting, heating and cooling, as well as communications, voice and digital.

That leaves the power segment to be discussed.

# INDUSTRIAL POWER SYSTEM AND FINAL ELEMENTS

Within the industrial power system there are three major categories of final elements, motor starters, solenoids (on valves of all sorts) and contactors (for heating, lighting, or other loads).

Again, there are graphic conventions, and they are intended to portray function not features, the symbols have no pictorial relevance to the actual product.

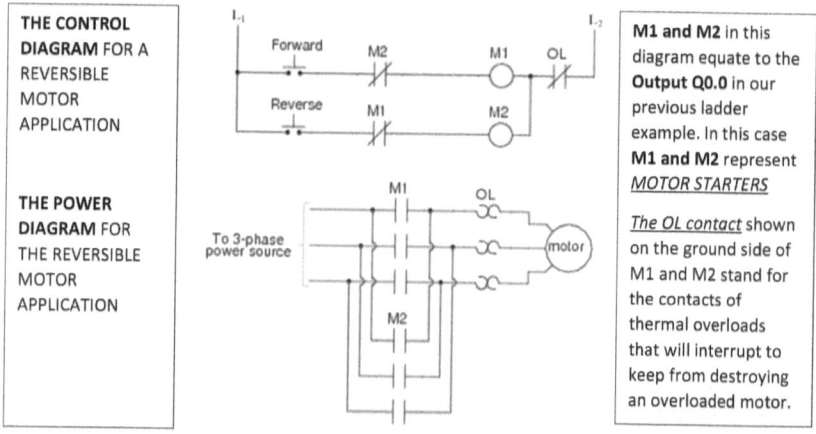

THE CONTROL DIAGRAM FOR A REVERSIBLE MOTOR APPLICATION

THE POWER DIAGRAM FOR THE REVERSIBLE MOTOR APPLICATION

M1 and M2 in this diagram equate to the Output Q0.0 in our previous ladder example. In this case M1 and M2 represent MOTOR STARTERS

The OL contact shown on the ground side of M1 and M2 stand for the contacts of thermal overloads that will interrupt to keep from destroying an overloaded motor.

This form of circuitry would be shown for every motor in the facility, and the related control would be shown in a manner similar to what his shown here or in the ladder we discussed earlier. So, that only leaves the solenoids.

To simplify the portrayal of the solenoid in ladder logic, we will readdress the first ladder we looked at, and add to it.

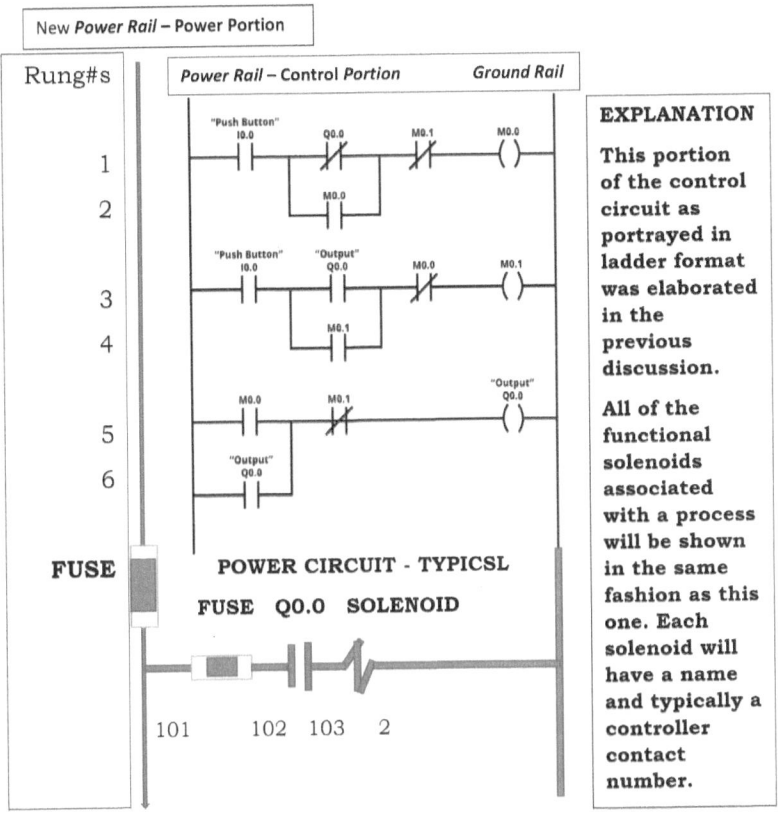

| Rung#s | Power Rail – Control Portion | Ground Rail | EXPLANATION |
|---|---|---|---|

This portion of the control circuit as portrayed in ladder format was elaborated in the previous discussion.

All of the functional solenoids associated with a process will be shown in the same fashion as this one. Each solenoid will have a name and typically a controller contact number.

So, we have considered all three aspects of industrial electrical utilization, and all three methods of diagramming the control of the FINAL (WORK) ELEMENTS.

I sincerely hope that if you get only one thing out of all the advice given in this book it will be an appreciation of the absolute need for, and value of internet research with regard to industrial processes and the means and methods of controlling and servicing the many systems encountered today.

Before we leave this aspect of our study, please review the following URLs and the variety of symbols used to discuss and describe

## ELECTRICAL

https://www.eaton.com/content/dam/eaton/products/electrical-circuit-protection/medium-voltage-vacuum-circuit-breakers/comparison-nema-iec-schematic-diagrams-mz081001en.pdf

## FLUID POWER

https://advancedfluidpowerinc.com/wp-content/uploads/2016/03/Fluid_Power_Symbols.pdf

## PLC *(Process controller)* LOGIC

https://www.plcacademy.com/ladder-logic-symbols/

By this time, you should be realizing that RELIABILITY CENTERED MANUFACTURING PROCESS MONITORING AND MAINTENANCE ACTIVITIES are critical to organizational performance and profitability.

In the next chapter, we will turn our attention to numeric systems of communication that are specific to the field of control and data management that many of us may use without knowing it, or knowing anything about it.

# CHAPTER 4

# THINGS DO NOT ALWAYS ADD UP

We have just concluded a rather elaborate discussion of ladder logic as it relates to the electric control and power applications in manufacturing.

Unfortunately, unless someone is working with ladder diagrams on a frequent basis, it can be a challenge to gain assurance and be capable of explaining the sequence of operations, and purpose of a logic circuit.

Another level of frustration is induced when a relay logic system is updated, and the method of control is transitioned to the Programable Logic Controller (PLC) generation.

Let us begin this chapter by considering the means of convincing yourself that a circuit works, and then enabling you to explain the circuit and its application to peers. We will use a tool identified as a truth table.

## THE TRUTH TABLE.

Too often when this type evaluation of performance is discussed, the publisher puts the circuit in the appendix, and discusses the truth table in current context. Since this is my book, I am going to violate editorial custom, and copy our former ladder circuit one more time.

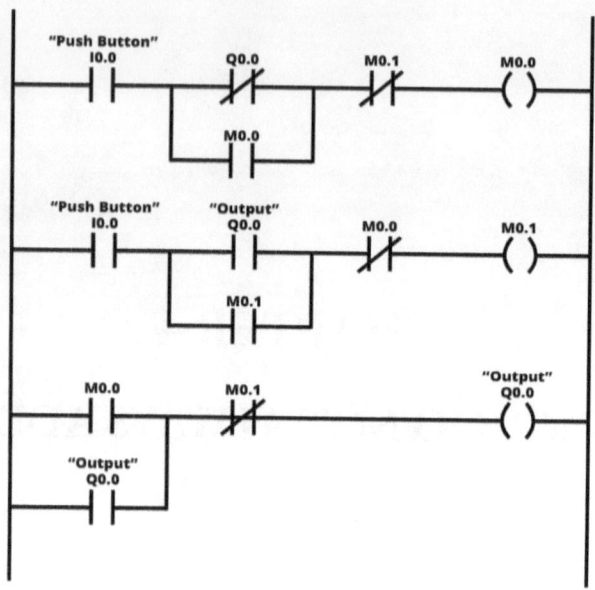

The first thing that we must do to begin to explain the circuit is to do what we would have already done. That being the developing of a **SEQUENCE OF OPERATIONS.** Such a document would have been developed on the basis of the process to be performed, and provided to the control designer to perfect this circuit.

This is an unusual industrial circuit, because a start-stop (toggle) circuit will normally incorporate two separate push buttons START and STOP, the start button being green, and the stop button being red.

This circuit is more like a residential lighting circuit with a push-on, push again-off. Or the car radio. It is identified as a toggle circuit. Yes and no from the same signal.

The *sequence statement* for the circuit shown however should go as follows.

1. Initial conditions Power is on, but there is no action.
2. Operator pushes I0.0 (the Input Push Button) one time
3. The Output Q0.0 is energized and the process is initiated.
4. The system will run until it is time to stop the process.
5. The operator pushes I0.0 (the same Input Push Button) one more time.

6. The Output Q0.0 is deenergized and the process is stopped.

7. The system should return to its initial condition.

As easy as this sequence is to read and understand, it would make more sense if we made the process do something. So, let us say that Q0.0 energizes a hydraulic pump motor to power a major fabrication and assembly process.

This would suggest that we change the following statements to read as follows.

3. The Output Q0.0 is energized to START THE HYDRAULIC PUMP MOTOR (Energize M1).

6. The Output Q0.0 is deenergized to STOP THE HYDRULIC PUMP MOTOR (Deenergize M1).

NOW TO PROVING THAT IT WORKS THE WAY IT IS DRAWN (TRUTH TABLE)

| STEP | I0.0 | M0.1 | M0.0 | Q0.0 | 1M |
|------|------|------|------|------|-----|
| 1 | 0 | 0 | 0 | 0 | 0 |
| 2 | 1 | 0 | 1 | 1 | 0 |
| 3 | 0 | 0 | 0 | 1 | 1 |
| 4 | 1 | 1 | 0 | 1 | 1 |
| 5 | 0 | 0 | 0 | 1 | 0 |
| 6 | 0 | 0 | 0 | 0 | 0 |
| CK | V | V | V | V | V |

The table shares the sequence steps and indicates the move of all the coils. In actuality, every contact should have its own column and prove each change of status and the return to initial conditions.

Experience has proven over and over that if it works on paper, it will work in practice.

# ONE THING ALWAYS LEADS TO ANOTHER

Subtitle That is what makes training effective ... PROGRESSIVE REVELATION!

What now?

In the truth table, we used 1s and 0s to illustrate yes and no; or on and off. This is a brief introduction to a numeric system known as digital. Digital is truly the language of the computer. But the computer is capable of multi-tasking with numeric data processing, so it integrates digital into registers and data storage tables in the forms of binary, octal, and hexadecimal; then precision encoders maintain positional knowledge by applying digital in a system identified as gray code.

The following is an example of the comparison between binary and gray code.

Binary to Gray Code Converter Table

| Decimal Number | Binary Code | Gray Code |
|---|---|---|
| 1 | 0001 | 0001 |
| 2 | 0010 | 0011 |
| 3 | 0011 | 0010 |
| 4 | 0100 | 0110 |

Although it appears that both systems are the same, a closer observation shows a subtle difference, in the Binary Code system, frequently, 2 or more digits change in the same step.

With regard to precision accuracy in monitoring motion by means of encoders and high speed digitally indexed data, binary and the other systems are totally incapable.

## NUMERIC SYSTEMS, BUT NOT MATH (ARITHMETIC)

We dedicated several chapters in our previous books illustrating that in math applications require the integration of numbers with words.

Either words that add descriptive information to the quantitative value of numbers, or clarity with regard the units of measure with which numbers are associated.

Numeric systems do not incorporate words, they are unitless. It is the arrangement of the numeric display and the universe of data represented that provides the descriptive data.

The reason for the use of the differing systems is based on the amount of data that can be managed in each. In terms of the data management capacity, the systems are arranged in the following order:

- Binary (Base 2)
- Octal (Base 8)
- Our common English System (Base 10)
- Hexadecimal (Base 16)

FOR THE FUN OF IT, CHECK THE FOLLOWING URL, entitled, "You put the Hex on me!"

**https://andybargh.com/hexadecimal/**

We will not attempt to dig any deeper into this topic, because these are the most modern categories of studies as they relate to computers and the means of programming computers.

As we wind this chapter down, I want to share a thought about pursuing information for the purpose of pursuing information.

The Bible shares that when much is given, much is expected, and when we know to do and do it not it is a sin.

We can pursue information incessantly, but the question that we all must face is what do we intend to do with what we already have.

One of the most challenging expressions I can recall in this regard is a song shared by a woman in her mid-thirties dying from cancer, and a paralytic also in his thirties as a duet began with the words;

*"With the time that you gave me, have I done all I could do?*
*Have I loved all I can love, and have I lived all I can live? …"*

I pray that this book, the website, the blogsite, and the 9 books that have preceded this one will find their way into the hearts and hands of readers that will build on the foundation this information offers.

In the following chapters, we are going to attempt to uncover the mystical layers like peeling an onion, as it relates to the disciplines that enable work, and their reliance on the control strategy and reliability for their performance.

The products that are available today are not the products that populate nearly 50% of our current manufacturers inventory of equipment.

The concern I am attempting to share is anchored in two primary issues. First, what we are now doing is in need of repurposing and repairing. To that population I have skewed this advice and counsel. Secondly, there is the current vow to bring major life sustaining manufacturing back to US soil from all corners of the earth.

The thing that is not being shared is that we helped move that manufacturing offshore in the first place, and we didn't ship old equipment. The skillsets needed involve the ability to manage, service and maintain the electro-mechanical generation of production equipment and processes. At the same time to be prepared to manage, service and maintain the mechatronic generation of equipment.

There is no question that a diligent student can gain a measure of workplace readiness with regard to the mechatronic generation, because like the aerospace automation, most of this style is equipped with self-diagnostics. The ability to monitor, diagnose, troubleshoot and repair does not rely as heavily on the human factor. *The need today is to know **what to do**, not know how to do what you are told to do.*

The current pandemic has brought about obvious proof that as knowledgeable as we are in the thrills, frills and frivolity that our American freedom has enabled, we do not have the skills, or the spiritual desire and commitment to home school our own children. We seem to be more willing to send them into the masses against the claims of health science, to sit in a classroom and hear a lecture on advanced science.

Why is there such a drought of basic math skills in America today? Why is it so difficult to write a 1000-word essay that makes sense? To vocally resist any suggestion that we attempt to write. Those that offer the greatest resistance are typically sharing many times more than that 1000-words via texting every day.

Why are new street signs symbolic instead of words? Words require the reading skill.

Why does a clerk in a fast food market struggle and call on a supervisor to make change for a $20.00 bill?

Is anyone really ready to believe that non-readers, non-writers, non-mathematicians will amass the money to buy the books, pay the fees, and commit to the time, travel and study schedule to succeed in a college diploma program. How many are willing, or capable of delaying applying for satisfactory work for the 2 to 4 years it takes to achieve a two-year diploma?

I know this all sounds like sour grapes, but I ask you to consider what I am truly offering in these books and the associated resources.

But like a progressive dinner, all of these are the appetizers, what I am truly offering is access to personal mentoring. Advice anchored in more than 68 years of putting my money where my mouth is. There is nothing in what I have shared or will share that I have not done, and cannot do now, given the handicap of age and agility.

My handicap now is based on physical deterioration. But the truth is if we can get together, we can share.

I may not offer to visit you, your site, or to assume a role of teaching a prolonged series of courses in a classroom. But with the web, and meeting sites available by zoom, go to meeting, google, and so many others, we will get together when you want to, and we can include those you want to include.

SO WHAT?

There's that question again. I have said what I have in these most recent pages to share that *TRAINING, IS ANOTHER OF THE LANGUAGES OF MANUFACTURING*. I have been a training specialist for more than 65 of my career years. I know you will not find many, if any at all will share similar views, but look around you, at your circle of acquaintances, your social media family, and run your own poll as to the basic needs.

I will make a claim, and then move on to more teaching.

The eight books that are available as e-books on the website

www.aimtrain.net

represent a cost of less than $50.00. But if every young person in your church youth group from the ninth grade and above were given this set of books on the provision that they read them and act on the advice given. We could effectively parallel the cultural norm in the European countries and train them in trade and craft tool concepts via webtrain initiatives before they graduated from high school.

Not everyone is college ready, or even college bound. The problem with the opportunities that are open to this population have been played down by family and friends as grunt work, and to be avoided at all cause.

Anything worthwhile is worth working for, and we have a lot of work to do, but our young people and the under-employed are worth the effort.

# CHAPTER 5

# THINGS THAT LOOK ALIKE
## – ... ARE NOT ALWAYS
## EQUAL OR EQUIVALENT

You would think that our discussion of numerics in the last chapter was sufficient talk about numbers. The truth of the matter is that the use of numbers, and the format of their use, illustrate the difference between the *descriptive* and the *action* words of numerics than any other topic. A close second, and still related to numbers, are the studies of hydraulics, and the sizing of conductors that we will see in a later chapter.

We will not attempt to address all of the nuances of the systems that are related to computer numerics. But we would encourage you to visit the following URL, and do some study on your own. Every number system is capable of displaying the same value, but the differing systems do so in a much more efficient manner.

**http://www.cs.bu.edu/courses/cs101b1/slides/**
**CS101.Lect11.BinaryNumbers.ppt.pdf**

Next, it is important to have access to a glossary of computer terms, to begin to speak their language.

**https://www.computerhope.com/jargon.htm**

The following is a good study on systems and programming in general.

**https://codeburst.io/intro-to-computer-science-terminology-f9ae64e75d27**

With all of that we still have not gotten to the root of the matter, the actual number systems and their use. The following URL gets us much closer, so visit it and gain an over view of the relationships between the systems, and then come back.

**https://www.tutorialspoint.com/computer fundamentals/computer_number_system.htm**

Honestly, these four websites offer enough of a challenge and adequate study to qualify for a semester of computer language studies in any college anywhere but the differences and the reason for their being the *system of choice* is still not clearly articulated. Until you understand the distinctions and can appreciate their value you are merely studying numbers again.

The truth of the matter is that most students that pursue computer numerics in a college setting are heading into career pursuits in computer design, programming, or advanced analysis with tons of data.

SIDEBAR. …

This is probably late in coming but it is crucial, and we will do all we can to promote its visibility and impact across all that we do. We are not attempting to tell you or show you how to do in any specific area that we cover. We are not in a position to KNOW IT ALL, and we are not attempting to transform you into a KNOW IT ALL.

What we are attempting to do in the many avenues of sharing is direct you to resources that are reliable and informative to the point that you can perform with integrity, without having to be a KNOW IT ALL.

The problem with education is the pace of change in the availability of information, and the absence of time and opportunity to prove theories and speculation. New textbooks are published every day, authored mainly by teachers that are required to publish to maintain currency of status in their credentials.

They publish what they believe to be true without ever applying it in actual real-world applications. In essence they are authoring their best guess. Academic textbook selection committees select optional texts, and pursue the newest offers, but they are not currently instructing, so they are not in a position to evaluate content reliability or relevance.

Next, the instructor teaches from the newly adopted textbook, and again has not personally experienced the application or its relevance, so the story goes on. As the story of the events in the garden of Eden goes, when something of a disruptive nature happens on a job,

- The student declares … that is what I was taught.
- The instructor declares … that is what was in the book, and I merely taught it.
- The selection committee defends by saying … the publisher recommended it.
- The publisher claims, we only print what we receive, the author must have been wrong.
- The author takes the stand and submits that the publisher altered the content to make it politically correct.
- And that is my story, and I am sticking to it.

I do not share this information from a think so position, specifically with regard to the author and printer/publisher relationship.

And I share this without names to illustrate the validity of the continuum I narrated.

In February 2020, I submitted a book of short stories, written by inmates in a state prison system over a ten-year period as these inmates studied in a Bible College that met multiple times each week inside the prison.

The curriculum was a three-year diploma program with access to graduate levels as time and grades qualified advanced studies.

One of the assignments for each individual was to perfect a story of the life and lifestyle, and general encounters that ultimately led to incarceration.

A number of the inmates took the assignments seriously, and saw the potential value of publishing their stories as a means of reaching young people in the free world with advice and warnings as to the consequences of inappropriate lifestyle decisions.

The college administration discussed this concept with the local chaplain, the local warden, and the department of correction administration, and gained the assurance that it would be permissible, and based on the increases in the incarcerated population, a great idea.

Bit by bit, and story by story the book was compiled over the entire ten-year period. The corona virus pandemic and the closing of the prison to volunteer visits for any programmatic reason gave us time to fine tune the book for publishing consideration.

Since February, this 300-page book has been reedited 5 times, each time taking out hints of locations, nicknames, street slang, and any shady inferences at the insistence of the publisher's editorial staff.

After the many hours of reediting, and accelerating frustration with the staff and their "political correctness mentality", I withdrew the book. I refused to have it published when it was now purely fictional. It read in many cases like a novel; no means of even any indication of true experience. More importantly the lifestyle change that led to the desire for the story to be told.

So, to bring this sidebar to a close and get back to the business at hand, I submit;

THIS WORK IS NOT INTENDED TO TELL YOU *HOW TO DO*...

THIS WORK IS INTENDED TO SHOW YOU WHERE TO FIND WHAT YOU NEED TO *LEARN HOW TO DO* ...

MOREOVER, THIS WORK IS INTENDED TO SHARE PERSONAL ACCESS TO THE AUTHOR AND THE AUTHOR'S EXPERIENCE AND LIBRARY OF RESOURCES. AND

*COACHING AND MENTORING SERVICES* FOR THOSE WHO WANT TO *EARN WHILE THEY LEARN* TO PERFORM IN A TRADE/CRAFT ENVIRONMENT.

## Summary

Because of the extended virus pandemic, educational institutions are modifying the entire educational experience. Study at home, or in some cases study part time in school or at home, but no full-time face to face in the classroom. Where we live, there are thousands of homes that will not have internet access before December 2020 or later.

Classroom participation and curriculum schedule and mastery is mandated by law, so the only alternative is homeschooling.

What Aimtrain is offering is optional, enrollment by choice, more than 1200 one-hour courses that prepare the student for productive manufacturing participation. These courses could be studied by students of any age but to be more practical, from the ninth grade up, along with required educational offers. Thus, producing a three to four-year head start on career preparation.

## ON THE ROAD AGAIN, AS THE COUNTRY MUSICIANS SING

Things that are difficult to draw from a study of the websites cited are system distinctives that set each system apart from our traditional decimal system.

System by system we will share some of these distinctives, but first, let's clear the air on some important matters.

In every numeric system, including the decimal system that we use

- *Zero* has a value (meaning as a digit)
- Numbers in these systems do not have units (words associated with them)

- In addition to the number having value, its place in an arrangement of numbers has value (this place value is dictated by the system selected)
- Although I have never heard of it being taught this way, I will submit that the array of numbers in any of these numeric systems takes on the appearance of a data spread sheet
    - Number sets are *displayed in rows*
    - the rows are *divided up like columns* on a spread sheet to establish places
    - Sets of these rows of data are identified as *REGISTERS*
    - Comparisons of the number sets in each system are not discussed as interpretations, they are engaged in as conversions, and the determination of equivalent values.
        - Every language system can display any value, they merely do it in a different form.

Back to another question, "WHAT YOU TALKIN' 'BOUT?" Let us address that question pictorially, by looking at our decimal system first.

Every set of numbers we will ever use is referenced to zero (0) as a starting point. The spread below is all there is to our decimal number system.

| LEVEL | 0s | 1s | 10s | 100s | 1,000s | 10,000s | 100,000s | 1,000,000s | 10,000,000s | 100,000,000s |
|---|---|---|---|---|---|---|---|---|---|---|
| digits | 0 | 1 | 2 | 3 | 4 | 5 | 6 | 7 | 8 | 9 |
| decimals | 0 | 0.1 | 0.2 | 0.3 | 0.4 | 0.5 | 0.6 | 0.7 | 0.8 | 0.9 |
| fractions (10ths) | 0 | 1 | 2 | 3 | 4 | 5 | 6 | 7 | 8 | 9 |
| ones column | x | | | | | | | | | |
| **whole numbers** | 0 | 1 | 2 | 3 | 4 | 5 | 6 | 7 | 8 | 9 |
| tens column | | X | | | | | | | | |
| | 0 | 1 | 2 | 3 | 4 | 5 | 6 | 7 | 8 | 9 |
| hundreds column | | | X | | | | | | | |
| | 0 | 1 | 2 | 3 | 4 | 5 | 6 | 7 | 8 | 9 |

So, as will be seen with a bit of study, we do not get any more numbers, we just reuse the same numbers over and over again as we add new levels.

One of the most important concepts is that every decimal that we ever incur exists between 0 and 1, or between any other set of whole numbers where ever they appear, In the same manner, every fraction that we ever incur exists between 0 and 1, or another set of whole numbers where ever they appear.

Lastly with regard to our decimal system and the portrayal we have just shared, if you hold this up to mirror, if you hold this spread sheet up to a mirror, everything on the negative side of zero is exactly as shown, except there will be minus signs in front of the numbers.

## WHAT THEN IS DIFFERENT ABOUT THE BINARY NUMBER SYSTEM?

By definition, Binary is defined by only two digits being utilized, zero (0) and one (1). And in a manner similar to decimal, the value of each digit is determined by its location in the display. Like the decimal number system every thing done on the plus side of zero is duplicated on the minus side. In binary, there are new terms introduced, nibbles, bits, bytes, and words. The explanation of these terms is clarified by visiting the following URL.

**https://learn.sparkfun.com/tutorials/binary/
bits-nibbles-and-bytes**

With that in mind, let us take a look at a binary array and the means of counting. All computer languages start at the right with the least significant bit, and read from right to left and top to bottom. The value of each column is a power of the base 2.

| THE BINARY SYSTEM AT WORK | | | | | | |
| --- | --- | --- | --- | --- | --- | --- |
| VALUE | 32 | 16 | 8 | 4 | 2 | 1 |
| 0 | 0 | 0 | 0 | 0 | 0 | 0 |
| 1 | 0 | 0 | 0 | 0 | 0 | 1 |
| 2 | 0 | 0 | 0 | 0 | 1 | 0 |
| 3 | 0 | 0 | 0 | 0 | 1 | 1 |
| 4 | 0 | 0 | 0 | 1 | 0 | 0 |
| 5 | 0 | 0 | 0 | 1 | 0 | 1 |
| 6 | 0 | 0 | 0 | 1 | 1 | 0 |
| 7 | 0 | 0 | 0 | 1 | 1 | 1 |

For what ever reason I was unable to add new columns here but the value of each number following to the point of reoccurrence is as follows; 8=001000; 9=001001; 10=001010; 11=001011; 12=001100; 13=001101; 14=001110; and 15=001111. At this point we add a column to the left with each entry being equal to 16. So, the number 16=010000; 17=010001 etc.

With binary the columns are added for every new set of 16 numbers, so the computer specialists offer another set of binary system known as the Binary Coded Decimal System. A visit to the following URL will provide a good description of BCD and its use. In essence, the four bits of BCD are the same as Binary, but the range of numbers is limited to 0 to 9.

Primarily the function of BCD is to power numeric displays. Please visit

**http://www.idc-online.com/technical references/ pdfs/electronic engineering/Binary Coded Decimal.pdf**

| BINARY | BCD |
|--------|------|
| 0 | 0000 |
| 1 | 0001 |
| 2 | 0010 |
| 3 | 0011 |
| 4 | 0100 |
| 5 | 0101 |
| 6 | 0110 |
| 7 | 0111 |
| 8 | 1000 |
| 9 | 1001 |

9

1 0 0 1

Each digital output will be displaying the numeric value of a set of four BCD digits as shown. BCD is limited to 0 to 9 as shown.

In the same manner, each thumb wheel data input set is served by a four-digit BCD data register.

There is a proliferation of data available via the web if you are still hungry for this level of data manipulation.

## WITH DECIMAL, BINARY AND BCD BEHIND US, WHAT ABOUT OCTAL, AND HEX?

Octal is based on numbers to the base 8. That means that the progression from the least significant bit on the right to the most significant bit on the left, each column added from right to left represents the next higher power of 8. So, the values in each box are powers of 8.

The following table illustrates the number ranges and identifies the base (radix) for the four basic computer languages. The key to which system is applied to which application is based to a great extent on the total amount of data that the numeric system must account for.

| Number System | Base | Digit Used |
|---------------|------|------------|
| Binary | 2 | 0, 1 |
| Octal | 8 | 0, 1, 2, 3, 4, 5, 6, 7 |
| Decimal | 10 | 0, 1, 2, 3, 4, 5, 6, 7, 8, 9 |
| Hexadecimal | 16 | 0, 1, 2, 3, 4, 5, 6, 7, 8, 9, A, B, C, D, E, F |

Suffice it to say, you have had access to more information on numeric systems here than the average business major in college is exposed to in a full four-year career. The reason we are covering this much detail in this manner is merely to raise awareness of the need, and offer advice and counsel for those that want to pursue some of those seven million unfilled jobs in skilled and semi-skilled positions in manufacturing.

For now, we will move on to investigate the language systems that distinguish the more pronounce skill needs in industry.

# CHAPTER 6

# MOTION MANAGEMENT ... THE MEANS AND METHODS OF CONTROL

In previous chapters, we discussed the components that provide the muscle to move tools and equipment to accomplish work. Now, we will take a look at the way in which the electrical system communicates with the final elements and how those final elements convert electrical and electronic signals to mechanical outputs.

One of the more dramatic changes and accelerated rate of change involves the transition from electro-mechanical switching that is discrete in its performance to electronic sensors that offer non-contact or analog signals.

The problem for the monitoring and service technician is that nearly 60 percent of the existing machinery in industry today is still equipped with discrete switching devices. At the same time, nearly 90 percent of the training available in a local environment overviews and deals with solid state electronic, and analog sensors.

Another category of great concern relates to the control system itself. Although there has been a great number of controls upfits from relay to PLC, there are still a great number of electro-mechanical relay systems in operation.

Recall, we discussed the categories of control as Information (Inputs), Decision, Output, and Final Elements. From the standpoint of the role of the technician, the broadest range of challenge is presented by the variation in design and function, the frequency of new product introductions, and market trends relates to the devices applied as inputs (command and feedback).

None of these systems were difficult when they were new, but years of operation, process changes, control upgrades, and the need to apply optional equipment on an emergency basis add complexity, confusion and ongoing concern.

Traditional electro-mechanical switches were typically two-wire or three-wire switches that were wired between a common voltage source (commonly rated at 115 volts ac) and the relay coil that contributed to the decision making of the system. The wiring to and from the switch was numbered as discussed with regard to ladder diagramming.

Unfortunately, although the switch may not have been replaced over the life of the machine, wiring changes are frequently made, and it takes an act of congress (a cliché) to update wire numbers on drawings. The failure to update the drawings in one way or another will ultimately cause grief.

Although switches are simple in design, and typically require very little service, wear and tear do produce the need for actuator adjustment. When repair or replacement do become necessary, replacement parts are typically in limited supply because of the age of the device, and service literature is seldom available. To add to the concern modern equipment is rarely interchangeable.

The most troubling problems with switches occur when a modification or an upgrade involves replacing an electro-mechanical switch with a solid state electronic non-contact switch (commonly called proximity switches).

There is an outside chance that the switch will operate on the same 115-volt AC power and utilizes the same two wire configuration. This is not typically the case.

A visit to the following URL will provide more than you will probably ever need to know, yet make you very comfortable, knowing enough to be safe while working with the more common devices.

**https://www.softnoze.com/glossary-s.cfm**

The most demanding challenges arise with a company determines to replace an old electro-mechanical relay control with a PLC system. Then added demands are induced if the decision involves transitioning from discrete switching to analog sensing.

The conclusion that one must draw with regard to the field of switching and sensing is that the information is prolific, the need for knowledgeable personnel is great, and the field will continue to grow.

The need is for the technician and those that desire to pursue the field to assume personal accountability for the knowledge base and research. This too must be pre-need. When a device fails, it is not the time to go to the computer to initiate research. However, it is better for the technician to go to the computer for a brief but effective research initiative, than to play it by ear and destroy the device or disrupt the program.

## COMPUTER RESEARCH

In the control's arena, and specifically with regard to switches and sensors and their role in providing the information for commands and feedback for an automated system, service literature will be readily available. Specific instruction with regard to application, installation, and connectivity will be more effective if sought as a course initially, and as a job aide or reference as a follow up to study.

As the "My Pillow" man suggests, "I interrupt this program to make an offer" ...

... a visit to ... www.aimtrain.net will serve as a catalog for every need you might have to prepare for employment in the workplace. We cannot guarantee successful employment, but we can assure you it will not be for lack of resources, or by reason of affordability, and if you make the commitment and do your part, for lack of effort.

## OPERATOR INTERFACE IN PROGRAM MANAGEMENT

Much of the aged equipment that is in current use utilizes operator workstations equipped with push buttons, selector switches, lights, and gauges. Today, the technology has jumped from a catalog of devices, to a touch screen panel with graphics, instructions, and in many instances, troubleshooting and diagnostic guides.

There are unlimited resources on the internet for your investigation and study. The first is an elaborate and detailed explanation of what a Human Interface Module (HIM) does.

**https://www.silabs.com/documents/public/ application-notes/AN249.pdf**

But there is a need to be enabled to see some of the popular options offered in the market today. The following URL will give you a good overview of the range of products that one major manufacturer offers. There are many manufacturers of such devices, and the market is growing.

**https://www.rockwellautomation.com/en-us/ products.html**

As the opening screen of this URL portrays, we have all seen examples of this type screen in our favorite fast food restaurant at the drink station. Another example of a domestic/commercial use is the cash register at the same fast food restaurant.

Then you may have met with one of these wonder workers in the local Department of Motor Vehicles when you went to renew your license.

## TRENDS

As astounding as Alexa is in our homes, cars, and businesses, accepting voice commands and performing the wizardry of control of radios, security systems, lighting, and door locks, industrial versions permit a business owner to run his business from home.

There is no arena of activity that experiences change at such an accelerated pace today than in the management and maintenance of automated processes. We talk so much about robotics, but virtually any new machine, or piece of equipment is capable of the motion patterns, accuracy and reliability of a robot.

All of the instrumentation that makes a robot a robot, can actually be applied to any machine. All it takes is money, and the knowledge of instrumented motion management at the time of the design and build. Aftermarket retrofits cost too much.

A visit to the following URL and diligent study will provide a basis for understanding the value add of instrumentation, with regard to the management of motion.

**https://www.omega.co.uk/techref/glossary.html**

Unfortunately, providing you with these glossaries is like opening a door and inviting you into a library, and suggesting you find the book you want. The degree to which you will be successful is dependent upon your understanding of the system of cataloging.

The libraries for example utilize a numeric system known as the Dewey Decimal System. A visit to the following URL will orient you to that system and its history.

**https://www.library.illinois.edu/infosci/research/ guides/dewey/**

In the library, when you are researching for a book, or a classification of books, the Dewey Decimal Number is the key to your search.

In industry, and particularly in researching categoric information pertaining to automation, control, power, instrumentation, your success is dependent on the search **KEYWORDs.** These are words that are common to the discipline or activity that you are attempting to research.

Too many times someone does extensive research about a topic using only the words of academia, while the specific topic they seek uses colloquial or street talk. Their searches will go down many rabbit trails. And they may never truly lock in on what they are seeking.

Visit the following URLs for a brief comparison of the difference in terminology, and then fasten your seat belt so we can take a ride.

**https://www.tolomatic.com/support/glossary**

This is a glossary of terms used by the motion control industry and product suppliers to that industry.

**https://www.motoman.com/en-us/about/company/robotics-glossary**

This is a glossary submitted by a manufacturer of robots for business, industry and the medical profession.

Comparing one of the first descriptions from each, let us see if their language is similar.

> **Accuracy (Manufacturers – Marketing vocabulary)**
> Accuracy is the measurement of the deviation between the *command characteristic* and the *attained characteristic (R15.05-2),* or *the precision with which a computed or calculated robot position can be attained. Accuracy is normally worse than* the arm's *repeatability.* Accuracy is not constant over the workspace, *due to the effect of link kinematics.*
>
> **Accuracy** (industry language) The degree to which an actuator is able to move to a specific commanded

point. On the bullseye below, notice that all the holes are centered around the middle of the target, but the grouping is not very close together. *Good accuracy does not require good repeatability.* **(see repeatability & accuracy)**

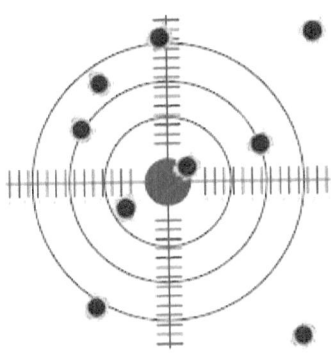

**Repeatability and accuracy** (industry language) The measure of how close to a programmed point the actuator can come, and how close it gets to that same point again. The repeatability of industrial actuators is usually much better than the accuracy. Notice on the bullseye below that the points are centered around the middle of the target and are grouped close together. *This is good accuracy and repeatability.*

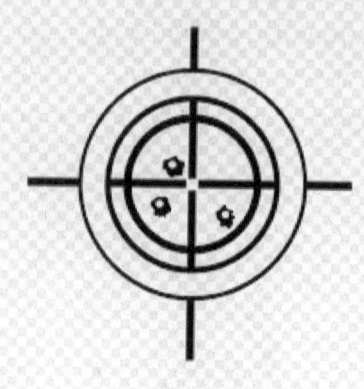

## THIS FIRST WORD COMPARISON ILLUSTRATES A CRITICAL CONCEPT

This is the ninth book, following hundreds of articles over these past many years in which I have attempted to illustrate in a believable manner, the distinction between EDUCATION and TRAINING.

It is true, that a trainee may take related instructional courses in an educational setting, and over a 2-year period averaging 12 to 15 class hours per week in class and lab (approximately 960 hours (+/-) studying the text book vocabulary and being instructed in theoretical.

In a parallel, coordinated OJT (monitored hands on continuum), over a period of 8000 (+/-) hour structured craft/trade related program this same trainee will be expected to translate that textbook vocabulary to street and workplace terminology.

Unlike the academic practice of grading performance on a scale, the trainee I the work environment is being evaluate on 100% performance mastery in accordance with national standards, and time-proven evidence-based practices. There is no scale or curve in performance, you either do it right, or you do it over, again and again until you do get it right.

A visit to the following URL will be enlightening with regard to the actual implication of terms in an applied sense.

**https://www.newport.com/n/control-theory -terminology**

## REMEMBER GARY COLEMAN'S QUESTION, "WHAT YOU TALKIN' 'BOUT?

In the next few discussions, we will attempt to address the implication of that question to motion management terminology.

We have already discussed ACCURACY, and REPEATABILITY, with regard to position management. But there are also the matters of ERROR, and TOLERANCE that also relate to motion management.

> **Error …** The difference between the actual response (of a robot) and a command issued.

> **Tolerance …** A specified allowance for error from a desired or measured quantity.

As important as these terms are to the technician and the effective management of motion, they are not deemed to be important enough to be included in the product manufacturers glossary. That may well be because they do not want to even imply that their products are capable of error.

## SOME SAMPLES OF WORKPLACE VOCABULARY CONCERNS

The desire of the industrialize world today is that all manufacturing process technicians (servicing and maintenance personnel) be multi-craft/cross skilled. That in itself is problematic from the cognitive standpoint, because the electrically inclined do not want to get oil and grease on their tools (perception), while the mechanically inclined, do not want to run the risk of shock (also perception – there is a procedure known as shutting the power off).

The welder does not have the desire to become an instrument technician, and the instrument technician has no desire to learn to do plumbing. Most of the reluctance is the dramatic differences in vocabularies, and more importantly the context in which their words are used. Add to that the perceptions we gain from TV about dirty jobs.

Like instrumenting a machine from the onset (design and build) the most effective cross training will be accomplished if we introduce and mature the trades involve simultaneously.

The need for such is expressed in two classic illustrations. Years ago, Paul Harvey a news analyst from Chicago, gained fame and notoriety by sharing the news as reported, and then shared *the rest of the story* as a well-researched and documented *back story*. That was the name of his show, and truth was his reason sharing in the way that he did.

Then at the movies in the 1970s and early 1980s we had the opportunity to experience, *"Fiddler on the Roof"* a narrative dealing with cultural upheaval in the life of a family and community. A father, seeing his daughters break with century old traditions inter-marry and go differing ways, frequently engaged in rhetoric with God, and after complaining would say, "On the other hand!"

Like the balance scale, there is weight on one pan, but on another pan, an equal amount of weight, is the rest of the story, yet a very dissimilar substance.

Possibly some of he most apparent contrasts in trade terminology occur when we attempt to cross-train *Electrical, Electronic, or Instrumentation* trades with mechanical trades, and particularly the category of mechanical that is known as *Fluid Power*.

The easiest way to share the differences

| CONCEPT | ELECTRICAL | FLUID POWER |
|---|---|---|
| POWER | KILOWATT HOURS | No meaningful equivalent |
| POWER input | 1 HP = 746 watts | 1 HP = 550 ft-lbs/second |

| | | |
|---|---|---|
| POWER output | 1 HP = 550 ft-lbs/ second, or 33,000 ft-lbs/minute | HP = $\frac{\text{psi x gpm}}{33{,}000}$ ft-lbs/minute |
| INTERFACE | MECHANICAL COUPLING | MECHANICAL COUPLING |
| OPERATION | MOTOR SHAFT (Rotates CW or CCW based on Wiring) | PUMP SHAFT (Rotates CCW or CW) Discussed as being opposite from motor |
| SAFETY DEVICES | FUSE or CIRCUIT BREAKER | RELIEF VALVE or RUPTURE DISC |
| SWITCHING DEVICES | SOLENOID OPERATED OR SOLID-STATE RELAY | SOLENOID OPERATED DIRECTIONAL VALVE |
| SWITCHING PATH | CONTACTS, discussed as NO (Normally Open), or NC (Normally Closed) | FLOW PATHS, discussed as NC (Normally Closed) or NO (Normally Open) *Opposite from the relay* |
| ILLUSTRATION | PUSH BUTTONS NO NC Not Flowing Flowing | VALVE POSITIONS (2-WAY) NO NC Flowing    Not Flowing |
| INTENSITY | VOLTAGE | PRESSURE |
| INTENSITY CHANGE | TRANSFORMERS | PRESSURE REGULATORS |
| QUANTITY | AMPERAGE | GALLONS PER MINUTE |
| QUANTIY CHANGE | RESISTORS, RHEOSTATS | FLOW CONTROLS OR VARIABLE PUMPS |
| CONDUCTORS | WIRE, CABLE or BUSS BAR | PIPE, HOSE OR TUBING |

| CONNECTORS | SCREW TERMINAL, or CABLE CLAMP | PIPE FITTINGS, HOSE FITTINGS or TUBE FITTINGS |
|---|---|---|
| MEDIA | ELECTRICITY | AIR, OIL or WATER |
| CONTAMINATION | ELECTRICAL NOISE, SHORTS AND GROUNDS | AMBIENT AND PROCESS DEBRIS AND WATER |

Before we move on to another topic, a visit to the following URLs will give you a feel for the importance of math in the trades.

**http://www.controlledmotion.com/fluidpowerformulas.html**

**http://acdist.com/wp-content/uploads/2014/02/Siemens-STEP-Series-Glossary-of-Electrical-Terms.pdf**

## THE MOST EFFECTIVE CROSS-TRAINING METHOD

Based on the way the disciplines are taught today, everything is linear, and the textbooks are linear. You study ELECTRICAL, OR MECHANICAL OR HYDRAULIC, OR WELDING ETC. That is the reason I am approaching you with the advice I am sharing in this book series and on my website.

The most effective cross-training method is to train all the disciplines that you are attempting to integrate concurrently, and match the curriculum content to relevant work activities. This introduces the concepts and promotes understanding through monitored participation and practice, in real world activities, on real world equipment, using real world tools and techniques, in real time, for real reasons.

Linearly, it takes a minimum of two years to accomplish a meaningful diploma program in Industrial Maintenance – Electrical Emphases, or Industrial Maintenance – Mechanical Emphasis. If you want to integrate the two disciplines, the colleges recommend finishing one, and then start over for the other. Of course, they will award advance standing for the core courses that are supposedly common.

The problems with this concept, by the time you finish one pursuit and start the second, it is at least two and possibly three years later. You will not remember what you studied three years ago, and you will not equate comparable concepts as relating to each other. The only time the electrician finds out he wired a motor backwards is when the other trade said I wanted that pump to run Clockwise, not your pump.

Both of them have been taught to narrate CW and CCW as they relate to the end of the shaft of their device. They are not taught that when you connect two devices you have to imagine you are moving the motor up to a mirror to make the connection.

To gain a real-life grasp of this concept. Take a selfie of yourself with your cell phone camera, but make certain you put something in your opposite hand and make sure you hold that hand so that it is in the picture. Which hand is in the picture? Which hand held the camera?

Better yet, walk up to a full-length mirror, and shake hands with yourself, is it common for you to shake with your right, and the other person shake with their left like the mirror is doing?

The biggest drawback for the student to such a concurrent strategy is the need for the basic tools of each trade, for each task. The biggest drawback for the employer/sponsor to this concurrent approach is the need for a mechanical and an electrical specialist to be made available to approve the performance.

## MORE REASONS FOR CONCURRENT STUDY

After the big stuff is covered, and the trainee is getting comfortable wit motors, and cylinders, and mounting patterns and mounting accessories, we get into the study of conductors and connectors. And, this point, pneumatics should be added to the mix, so that the ratings and specifications for each trade are on the table all at once.

Electrical trades and professionals are seemingly all aware that electrical conductors and connectors are sized for their ability to carry amperes. The rating is discussed as ampacity. For the most part, electrical conductor ampacity is determined by the material the conductor is made of, and the size (diameter or gauge) of the conductor.

Voltage compatibility is not a concern with the conductor itself, but a major concern when it comes to the insulation with which the conductor is supplied.

Connections for electrical are generally made with screw terminals, wire clamps, or splices, and splices are commonly soldered for cohesiveness.

While considering the electrical conductors and connections to the various components of the hydraulic and pneumatic system, it is the best time to investigate the conductors and connectors in each of these systems.

Whereas electrical conductors are nearly always wire-forms, either solid wire, or stranded, and most often copper or aluminum and the various insulation applications are applied as coatings.

You wouldn't think of wires wearing a hole in themselves and shorting out, but I had occasion in my foundry experience to watch a welder repairing a broken member on a machine. He was in such a hurry to get to the job and get it repaired, to avoid excessive downtime charges, that he only uncoiled enough of his welding leads to reach the job.

As he struck an arc and welded the broken member the coiled loop of cables on his welding machine expanded and contracted in the same manner as a garden hose on a hose real as you open and close the nozzle. Before he began to as done welding, I saw sparking between one of the cables and the machine. I called his attention to the matter, and he would have argued with me if I had not been his supervisor.

His position was that sparking was not possible, because the machine frame itself was not grounded. I insisted we take his machine to the shop and uncoil all of the cables, both the welding cable and the ground. We found cracks in the insulation in many places, and exposed copper in several spots. The cables themselves had scuffed the insulation against the frame over prolonged periods of use without the cabling being uncoiled before use. The truth is that the welding cables are rubber covered stranded copper wire. The heat and electrical load cause the wires to uncoil slightly under load expanding the cover, and scuffing against each other or the frame.

Since hydraulic and pneumatic conductors are frequently hoses, to accommodate flexing and traveling, they must by securely anchored, because they flex and move under pressure as well as travel with moving tooling.

The greatest problems with pipe, hose and tubing is misapplication when transitioning from one type connector to another. Each of these conductors is marketed on the basis of inch or metric sizes. But, 1 inch does not mean the same for a pipe, a hose, or a tube.

A visit to the following URL will give you a good overview of the marketing specifics for these conductors.

https://pipeandhose.com/book/pipe-vs-tubing

The issue with these conductors is not the descriptive dimensions that differ from pipe (marketed on the basis of the inside diameter dimension), tube (marketed on the basis of the outside diameter dimension), and hose (marketed on the basis of the inside diameter dimension).

The performance concerns are impacted by the means of manufacturing, and the inside wall texture. Tubing which is smooth walled can be pushed to a flow rate of 30 feet per second under high pressure. Pipe which is generally cast has a rough texture and should not be pushed to flow rates above 15 feet per second on pressure lines.

And hose, which generally has a rubber lining with an orange peel or golf ball dimpled lining should not be pushed above 8 feet per second. These flow restrictions and the difference in determining their inside diameter impact the size selections when transitions are made.

As a general rule, I have always advocated that when going from pipe to tubing, you should jump up one size for the tubing and fittings. When going from pipe to hose, jump two sizes for the hose and fittings.

Many of you may take me on by doing the math and showing that based on normal wall thicknesses and pressure ratings that what I recommend is an overkill. If the conductors are all you consider, I will agree.

But please before you react too violently, consider the connectors, and the broad variation which they are made. Some forged, some machined, some cast, and many differing styles of conductor to connector make up. All too often, the designer fails to take in the pressure drop of fittings.

Conductors and connectors would be a challenging study and exercise for someone that had been entrenched in electrical work for a number of years. Their normal response is selecting a conductor shouldn't be that difficult. In electricity, if I want to go from conduit to flex, I connect them at a common junction box, and run the same wire.

What distinctions are challenging for the student being cross trained into pneumatics and hydraulics? Try terms like

- Compressibility
- Incompressibility
- Viscosity
- Viscosity Index
- Free air ratings for portable pneumatic tools

We will let you look those up on your own, going back and revisiting a URL we saw earlier.

**http://www.controlledmotion.com/ fluidpowerformulas.html**

There are so many more things that could be shared but unless you are in the training, they are not likely to provide any benefit beyond the nice to know.

# CHAPTER SUMMARY

I am always offended when someone says there is so much more that could be said on this matter, but we will close for now and address it at another time.

We are not going to do that. If this book has sparked some interest on your part, you will find a virtually open-ended resource at

**www.aimtrain.net**
**and**
**www.mpsharecare.com**

| | |
|---|---|
| Challenge videos | Online E-Books/Power Points |
| Radio Interviews | Our Blog |
| Skill-based Knowledge Transfer | Mechanical/Electrical |
| Services to Adults | Human Resources |
| Ongoing Mentoring/Coaching | 5-Minute (Lunch and Learn) |
| ONLINE CLASSES Study at home | *Course lists awaiting your study* |
| Your own time, 24/7, (Libraries) | |
| Safety Health, and Plant Science | |

# CHAPTER 7

# A CHALLENGE FOR ORGANIZATIONAL CHANGE AGENCY

## DOING DIFFERING THINGS – FOR PROFITABILITY

An ancient adage shares, "Give a man a fish, and you will feed him for a day; but teach a man to fish and you will feed him for a lifetime". (Source uncertain)

That can readily be applied to all that I have said thus far, and what I am about to say. Related to job performance in the workplace, you can teach the task requirements of one job by means of repetitive OJT, but developing a career skillset involves much more.

The workplace today is far different than it was 60 years ago. We have seen the need for flexibility and nimble management this past 6 months in repurposing a manufacturing line from autos to respirators. A clothing line to masks and personal protective equipment for first responders and health care providers. And military vessels and field paraphernalia to hospitals and triage for virus victims.

American ingenuity is unparalleled in the world, but so much of that talent is possessed in the upper levels of a company's payroll.

Workers at the shop floor level and front-line supervisors are bored and ready to change employment for a dime an hour or better benefits. They do not feel challenged by their current task-oriented assignments, and they are not impressed by menial displays of employer loyalty or respect.

I have written this entire series of books to address opportunities for the unemployed. But truthfully the greater gain is available to the underemployed. And in many ways, we are all underemployed. Scientists and medical researchers have proven that none of us use more than 10% of our brain power on a routine basis. And if we are honest with ourselves and those around us, everyone of us is capable of more than we are currently contributing to make this a better place to live.

America's problem is attitude. Rioters and looters on the streets of our city's night after night feel that they are entitled to do what they do. Young adults and young marrieds run up credit card debt attempting to start their adult life at the level their parents are after 30 or 40 years of struggle. High school graduates enter colleges and universities with no sense of career direction, and after six or seven years of seeking, return home, saddled with many thousands of dollars of educational debt.

Parents and career counselors advise the young to avoid dirty jobs, and involvement in the sweat shops of major manufacturing. Government career centers test for the best college diploma or degree program, and Industry offers tuition reimbursement for a B grade or better in a college curriculum program.

That is not the way it should be. Unfortunately, the underemployed at the bottom of the food chain is virtually helpless to change the outcome of his future on his or her own.

The one advantage the underemployed have over the truly unemployed is the work and income they do have. In the pages of the last 7 books and much of this one so far, we have shared a means of self-development, and by means of a web site available at URL. ... www. aimtrain.net ... we have shared the titles of more than 1000 online courses.

Beyond that, we have shared our contact information for coaching and mentoring *to enable someone to begin* remediating academic weaknesses one course at a time, or in convenient bundles.

The keywords here are *to enable someone to begin*. The average semi-skilled program of career preparation takes approximately two years and involves about 300 hours of sequenced craft/trade related studies, integrated with 4000 hours of supervised, mentored, and evaluated work experience guided national performance standards. Semi-skilled programs from four to six years, with 600 to 750 hours of studies and from 8000 to 12000 hours of directed work experience.

Someone that is struggling to make ends meet because of under-employment or unsatisfactory employment may be able to begin, but the very magnitude of the pursuit is overwhelming if all they can accomplish is one course at a time. Advisors share that something is better than nothing, but most of us have read enough of the Bible to realize that the lifespan of humans was shortened shortly after the "Great Flood", approximately 5000 years ago.

THIS SOUNDS SO NEGATIVE … AND IT IS.

ANOTHER ADAGE DECLARES, *"IF YOU WANT DIFFERENT OUTCOMES, DO DIFFERENT THINGS"*.

## SO, LET'S LOOK AT A PLAN OF ACTION.

Based on a 68-year career of promoting skill training initiatives, I submit without fear of contradiction that a person that has no dog in a fight, will seldom if ever stay the course. What I mean by that is that the student needs to be invested personally, to feel the need to stay the course.

With interest rates at their current levels it is not feasible to borrow the necessary enrollment fees, but a resurrection of two employee benefit programs of the 50's to 90's would make this initiative doable.

**FIRST.** The payroll deduction option. Company pays for the enrollment of each student (Via the website and courses listings … all that a student could possibly study and work full time for less than $650 per year), and recover that through payroll deduction (approximately $55 per month, or $12.50 per week.

**SECOND.** Tuition reimbursement upon quarterly or semi-annually submitted grade transcripts with a grade average of a B or better, and a proportionate number of class completions. If a student enrolls in 50 courses for a year, then tuition reimbursement for a quarter would require 13 course completions.

**THIRD.** Align Job Descriptions for Semi-skilled to Skilled, to Semi-Professional positions in one-year increments and set up progressive training modules from entry level to initial salary positions in a manner similar to the following.

- Entry level to OPERATOR
- Operator to SET-UP OPERATOR, UTILITUY OPERTOR
- Set-up operator to GENERAL MAINTENANCE MECHANIC
- General maintenance mechanic to MACHINERY MECHANIC
    - to MAINTENANCE ELECTRICIAN
- Machinery mechanic to TOOL AND DIE, QUALITY OR TRIBOLOGIST
- Maintenance electrician to INSTRUMANTATION TECHNICIAN
- CROSS TRAIN maintenance mechanic to MAINTENANCE ELECTRICIAN
    - Maintenance electrician to MAINTENANCE MECHANIC
- Tool & Die, Maint. Mech., Maint. Elect etc. to SALARY


## FOURTH.

- Determine what you typically pay a temp agency for a candidate (Hourly rate)
- Determine what the temp agency pays the employee
- Split the difference and start the employee on a probationary program that involves academic assessment, and remedial studies to be accomplished in the first six months.

- Determine the prevailing pay scale for journey level Tool & Die, Mechatronics Technician, Instrumentation Technician, Quality technician, or Tribologist (Lubrication and oil analysis specialist).
- Split the average range into 8 or 10 equal segments for 6-month progressive wage increases, and start the program.

## FIFTH.

- The most it can cost you in out of pocket that you would not have in the event of a temp turnover is the enrollment outlay, and that only the portion not yet withheld in payroll deduction.
- What you would gain is employee loyalty, rooted in a feeling of worth, with less willingness to jump at an offer of a dime, and job training.
- Consistent technical vocabulary and training across the entire organization.
- Enhanced morale, reduced tardiness and absenteeism, and fewer personnel grievances due to assignment misunderstandings, and performance standards to revert to in the event of questionable performance.
- A skilled workforce, in readiness for advancement needs, at a pace more demanding than expected.
- Organizational sustainability, every employee knowing their job, the job they left, and the job they are working toward.
- Budget stability,

## SUMMARY

There are very few still around to remind us of the days when the skills were attractive, and pride in workmanship distinguished the craftsman. That is why I have been writing and offering this series of books as a challenge by which to "Make America Great Again" by making Americans Great Again, one at a time.

Rather than another government program, driven from the top down and resisted by those that view it as intrusion, let's make it a grass roots program and grow it from the bottom up.

There was a program offered by a training company in the early 1990's focusing on total employee involvement. It was a good program, well designed and uniquely focused on the mutual benefits if the employee adopted the attitude of ownership.

Unfortunately, it was about ten years too late. Industrial leadership, educational gurus, and the press had persuaded parents to direct their youth away from, "Those dirty factory jobs". The concept that was never shared as a contrast is that with soap and water, and a little effort the craftsman with pride in workmanship and a wholesome attitude can *go home at night with clean hands, and sleep at night with a clear conscience.*

We have a small window (two to three years) that will be greatly impacted by the pending 2020 election, to make meaningful steps toward employee retention and improved human relationships. As foreign owned business build on US soil, they are rifle shooting our workers, and when we bring major businesses back from offshore, it is only going to get worse.

You have heard of the survival of the fittest? Let me assure you, that relates to much more than competition in the marketplace, or competition in the sports arena. Today, it equates to supremacy in the Human Resource Administration.

Unfortunately, these are not the topics addressed in an MBA curriculum. The narrow focus there is on money, by any means. Nearly 35-years ago, while we were selling Apprenticeship related instructional materials in booklets that represented 10 hours of study each, the first book in nearly ever lesson plan was a book entitled, "How our Economic System Works!".

The emphasis was on the value added by each and every individual in a productive organization. Today, it seems to be attributable only to the total productive efforts of manpower and machines.

We need to return to the belief that everyone is essential, emphasize in one on one evaluations and coaching initiatives that, "Nothing works if every individual doesn't". I used to tell my children, you don't have to do it all, but you must do what you can do, while I do what I can do, as long as we all do our best, and we all work till the task is completed.

- **THINK ON THESE THINGS**
- **TRAINING** is not an expense.
- **TRAINING** is an investment.
    - Not in the individual you train.
    - But in the future and stability of your company.
- **TRAINING** is not analogous to CAPITAL INVESTMENTS
- **CAPITAL INVESTENTS** usually produce a 15-year life expectancy
- **THE TRAINEE** should be productive for a working career
- **CAPITAL INVESTMENTS** begin to decay immediately
- **THE APPROPRIATELY TRAINED CRAFTSPERSON**
    - Should engage in self-directed continuous improvement
    - Will employ valuable product and service research techniques
    - Will be a self-motivated lifelong learner.
    - Will not need to incur educational debt
    - Will have a four to six-year jump on a college graduate regarding seniority and organizational trust and confidence
    - And the list could go on and on.

I close with this thought. Each of us have choices to make with regard to what we know, what we are willing to do with what we know, and what and how we choose to share.

I have chosen to share what God gave me through my 86 years of living, with 68 of those years being devoted to studying about and mastering the means of training multi-craft skills that are so desperately needed in manufacturing.

Now, it is up to you the reader either as a student, or a sponsor for such to decide what you will do with what is offered here.

You can choose to benefit from what I have learned, and shared. Or, you can choose to make your own mistakes, and possibly live long enough to write about them.

But remember, *no choice is a choice.*

In that same vein of thought, no vote is a vote, so if you don't choose to vote, don't complain, and most of all go to the streets and loot, riot, and cry that someone has been unfair to you.

Nothing in the Bible, or Biblical history says that we owe anyone anything beyond emergency provision. No one is against a handout in the event of a genuine need, but no one should be made to feel obligated to provide a handout that is demanded out of disrespect.

If anyone is healthy enough to hit the streets and demonstrate night after night, they are healthy enough to work, and provide for themselves.

### CLOSING THOUGHT

**JESUS SAID, "By this shall all men know that ye are my disciples, that you love one another".**

**The nightly news leads one to ask, does Jesus have any disciples left in this day and hour?**

# CHAPTER 8

# PUTTING IT ALL TOGETHER

## FOR THE GOOD OF THE CAUSE –

Truthfully, it has been quite a journey, precipitated by the Covid 19 pandemic of 2020. At the earliest awareness of the virus, our visits to the prison for Bible College Activities were temporarily terminated. Temporarily in February, like I went on midnight shift temporarily for two years in my early ventures in the foundry.

My first thoughts while contemplating quarantine were to finish a book on the prison bible college, and student stories gathered over the 10+ years. But we had been entering and editing those stories as we went along, so there was not much left to prepare them for publishing.

While my wife made several more editorial paths through the prison bible stories book, I began to jot down some thoughts about a book that would deal with the problem of unemployment, under-employment, and the issue of schools closing in favor of online instruction.

Before long, my notes became almost unwieldy, so I began to key in my thoughts and in early March, I sent a manuscript entitled, *"Nothing works, if you aren't willing to"* to a publisher, along with the manuscript for the Prison Bible College Stories manuscript.

The publishing process seems to drag on forever, with drafts, permissions, and editorial changes, but by the end of June, "Nothing Works…" had been printed and was in the earliest stages of marketing.

The Prison Bible College Stories met with editorial reluctance on the part of the publisher. They insisted that we get individual letters of permission for each of the studies included, and many of the authors had been released, or died over that period, and with the virus quarantine, we could not gain access to the few that might still be incarcerated locally.

Once the creative juices started to flow, and the thoughts of sharing the experience of my 68-year career working in and training workers in the area of semi-skilled (operational positions) and skilled (servicing and maintenance positions) performance as evaluated against (objective) national standards as opposed to local (subjective) learning objectives, motivated me to keep on keeping on.

Giving thought to the scope of all that we might share, and how to relate it to someone that did not have the benefit of similar experience, I decided to offer the differing topics as they relate to the biological systems of the human anatomy. I even toyed with the idea of entitling the series ... *The Anatomy of Automation.*

Having the first book already in print, and available through Amazon as an E-book, I determined it would be better to build on the theme that I set forth in *Nothing Works.* I wrestled with a few titles for the series, and settled on the theme, ***"If it is going to be, it is up to me".*** Feeling that if I could motivate anyone to explore training, possibly I could motivate them further. Assuming the initiative to pursue a self-directed action plan. And possibly pursuing a credentialled skill training course and become a recognized tradesperson in either mechanical, electrical or lubrication technologies.

Every time the publisher insisted that I water down the Prison Bible Stories book, it heightened my frustration, it was now no longer a biographical history, it was becoming a novel. A work of fiction, that lost its central thread of continuity; the redemptive response of a loving savior, when an errant runaway gets *sick and tired of being sick and tired* and cries out for cries out for redemption and restoration.

By mid-July, the frustration had grown to the point of righteous indignation (that is a Christian's definition of anger) and I turned to a recent parolee that had been a computer systems specialist and army recruiter nearly 20 years ago. I asked him what would it take to put together a website, and put my work out to the public in digital format and social networking.

A week later, he called me and asked me to log in and preview one of the most astounding websites I have had the privilege of viewing. Visit www.aimtrain.net and see if you agree with me.

While there, visit the blog, the online courses, the e-books and power points, and the services for adults' pages. Then investigate the Blogsite that is shown as a separate resource, and our other site.

Each of the books is offered on the website as e-books. I am not a man of means, and the publishing fees on the first book would doom this entire work to dormancy if they went the way of the first book. My hope is that someone will find this website, or blog site, see its value and help publicize it.

For the rest of this chapter, I will give you a brief overview of the preceding books, so you can determine which if any you would like to purchase and use.

## APPLICATIONS FOR THIS BOOK SERIES

For an investment of less than $50.00, a junior high school counselor could possess a full set of these books, and lend them out to students that were in a quandary as to what to pursue in high school, or as a career.

For the same investment a parent could provide fact-based advice and counsel to their children, or to an entire neighborhood.

For the same investment, a church youth group leader could offer the advice and counsel of a professional skills coach and contextualize it in a faith-based presentation.

That same youth group leader and his sheep could literally go to the streets and share the opportunities for career coaching classes at their church. The youth and their families might think about offering snacks. The church might even offer the opportunities of a hot shower, and coffee shop atmosphere once or twice a week I they were that evangelistic. Think about this, if these that are encouraged in such a manner find work and get established, they need a church home. *The Bible says to the **believer**, ... **GO YE!***

It is not out of the question for an instructor to invest the $50.00 and perfect his or her own curriculum to direct non-professional students into meaningful pursuits in business and industry, and there are none more rewarding than reliability services for manufacturing processes and automation.

I just pray that these books and the web offers do not end up being like a note in a bottle, or a time capsule that takes years to find a distant audience or unfamiliar population. The need is now.

As I begin to address the book series, let me share a picture that illustrates the progressive revelation process of a structured program as discussed in the previous chapter. Whether you follow this structure for one skillset, or pursue the integration of two or more skillsets, this works.

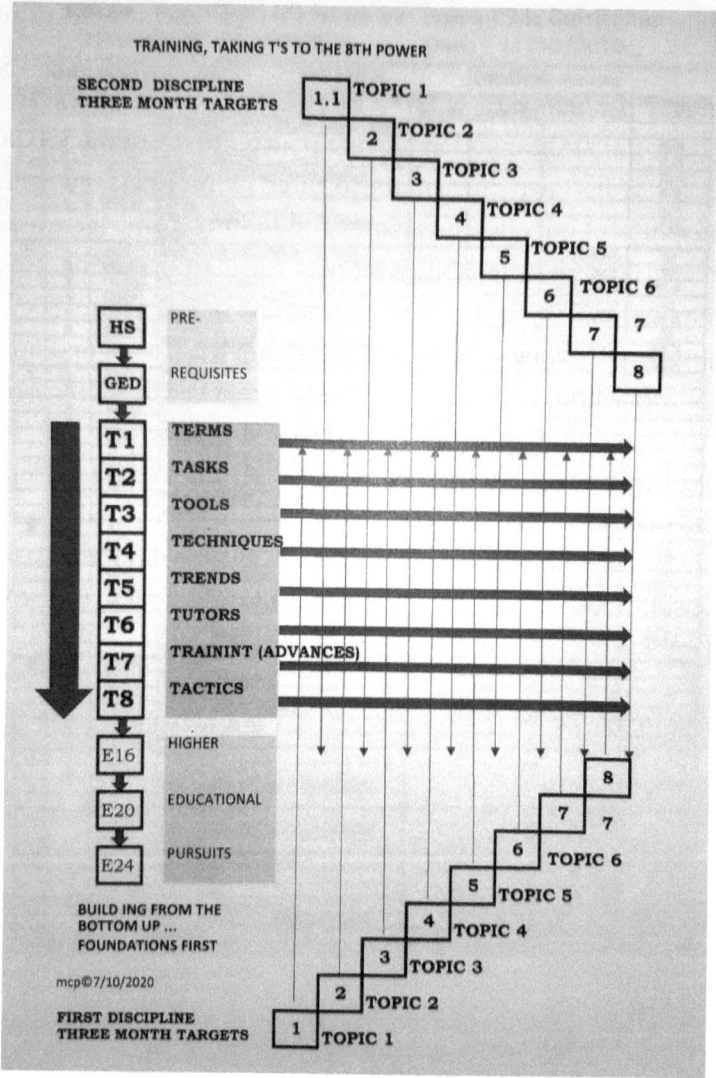

No matter how you approach it, YOU MUST begin with vocabularies, and not merely words, but words and numbers in context that are essential to communication in that specific discipling.

Then YOU MUST identify tasks that uniquely distinguish the skillset you are working toward from other similar skillsets.

Thirdly YOU MUST identify the tools, their proper selection, use, care and storage. TOOLS truly make the living. The *tool user* merely goes along as the chauffer, or caddy.

Fourthly YOU MUST identify the source or resource for proper tool techniques. A mechanic may properly select a torque wrench from its place of storage, equip it with the appropriate socket, run a nut down finger tight as instructed, and use the torque wrench in a correct manner. But if that mechanic does not follow the appropriate torqueing pattern (technique) for the specific joint. The process may lead to joint failure.

Everyone in the field today MUST BE AWARE OF AND IN TUNE WITH TRENDS, as they relate to TERMS, TASKS, TOOLS, and TECHNIQUES. When new products, or new equipment comes in to the work environment, the mechanic, or technician by any name, MUST seek TUTORS, OR TRAINING, OR BOTH as to distinctions and use and servicing mandates. Merely swapping out products that look alike can destroy not only the new piece, but the system it is installed into.

Lastly, the performer will by nature develop TACTICS that are unique to his or her personal value add. They may be attitudinal, or they may relate to a special selection of a tool or technique that works better for them than the published manner.

Consider Tiger Wood's golf activities during 2020 and the concern he has expressed attempting to use a new putter. Sometimes it is more productive to stay with old tools, old practices, and the advice of old timers.

With all of that having been said, I will conclude this brief narrative by sharing that YOU MUST go through these 8 steps with every new skill concept introduced. Terms, Tasks, Tool, Techniques, Trends, Tudors, Training (update and advanced) and Tactics.

## NOTHING WORKS: IF YOU ARE NOT WILLING TO

The seed that was planted, and the very excavation in which the footing for *IF IT IS GOING TO BE, IT IS UP TO ME* was laid; is a bit more of a pep rally than the actual game itself.

It has some of the coaching psychology that was commonplace in the mid 1900's. Think about some of the more popular motivational speakers you have heard about, Rockne (Notre Dame), Lombardi (Green Bay Packers), Knowle (Pittsburgh Steelers) and more recently, Lou Holtz.

Years ago, a plaque above the doors of the training center in Southern Indiana read … *Where the **will to achieve** is met with the ability to do.*

We gave quite a bit of attention to willingness in this first book. Skills are not caught, they are taught, but if the student is not willing to be a student and exercise the patience to be obedient both to the task and the Tutor, the process will not be a success.

**NOTHING WORKS … If you aren't willing to!**
*A Self-help for unemployed and under-employed.*

## TRUTH OR CONSEQUENCES!

If we are on the same page and you are curious about what is being said about the American Workplace on the presidential campaign trail, or in the evening news, you are in for a thrill ride.

Much is said about unfilled semi-skilled and skilled jobs; currently reporting more than 7.2 million jobs unfilled. At the same time there are approximately 6.9 million unemployed or under-employed. Who knows how many have quit looking by virtue of continual turndowns?

Nothing, however, is being said about the near total absence of personal performance skills; knowledge, abilities, aptitudes, attitudes and the like as being the reason for the turndowns.

Think about it. On the same newscast, you hear about the national dilemma posed by college debt. Someone with many thousands of dollars of college debt is reluctant to accept employment that renders them severely underemployed (not making a wage that lets them meet basic needs and start to retire the college debt).

Surely there must be a way that a serious applicant can find employment that satisfies their entry level needs, while meeting the employer's expectations at the same time. The good news is, **THERE IS!** The bad news is, it is not being reported on, or promoted.

Think about this performance skill list. Consider each variable individually,

- Knowledge
- Abilities
- Aptitude
- Attitude
- And the like
    - Personality
    - Self-confidence
    - Self-motivated
    - Self-disciplined
    - Fiscally mature
    - Communication skills
    - Inter-personal relationship skills
    - Demeanor including grooming and dress

The first three of these qualities are typically evaluated by potential employers, or recruiters. These three qualities relate to the value you can add by virtue of what you have done, and what you can do.

The rest of this list is flavoring and relate to who you are; and who you are willing to become given the opportunity.

Unfortunately, too many applicants are completely unaware of these differences, and they try to sell themselves based on "history", the past. The resume should provide all that is needed regarding history. Over the many years of training, we have coined a word for this type person, "US'-TA-WAZZERS"!

To expand on these qualities a bit more, the first three are taught. From birth, to death, we are perpetual students of life. Life happens, and although we say we learn through educational pursuits, the truth is we learn through experience, validating or disproving educational suppositions.

The rest of the list may be modified and flavored by life, but they are inherent. They are who we are. They are the foundation on which we build our life here and now.

### "AS WE BEGIN" the Author's challenge!

Let me share at the onset that I am an extremely positive person. I recommend that you take what is being said by the press, and our national leadership with a grain of salt. Nothing could possibly be that far off track. The news repetitively declares that the millions that are unemployed do not have the skills to fill the several millions of available jobs.

The truth of the matter is that those that are unemployed or underemployed have not been offered the opportunity to acquire needed skills. The purpose of this book is to share the failure of our educational system, and the absence of training alternatives.

In the pages of this book we will share a history of success and evidence-based methods of skill transfer. You will be totally wrong and willfully ignorant of EVIDENCE BASED FACTS if you read this, and don't challenge, or join hands and assist in INITIATIVES OF IMPROVEMENT.

I am probably the most POSITIVE person you will ever meet when it comes to FAITH IN WHAT COULD BE. But to get to what could be, you must first admit that you are sick and tired of being sick and tired of meaningless pursuits, and CHOOSE TO OPERATE the machines (our human lives) in the manner (life and lifestyle) that the DESIGNER (THE CREATOR, GOD) intended.

**This is not intended to be a RELIGIOUS BOOK, or to follow a RELIGIOUS THEME,** RELIGION has become like KUDZU after the 1939 WORLD'S FAIR, it was introduced as a cute little ornamental plant, and now, you cannot travel anywhere in the Southeastern United States without seeing mile after mile of forestation engulfed in vines that have no functional value.

Through Centuries, RELIGION has grown out of FAMILY AND TRIBAL CULTURES, through the customs, traditions, and acquisitions of PEOPLE GROUPS, then to A UNIVERSALLY ACCEPTED BELIEF SYSTEM (2000 YEARS AGO). SPLINTERED most recently by 1000'S of DENOMINATIONS, scattered in the wind like spider webs.

The recipe for most of these offshoots seems to be basic (intolerance and unforgiveness among the peoples) but the seasoning (PHILOSOPHY, PSYCHOLOGY, CUSTOM, TRADITION, FAMILY OR LEADERSHIP DOMINANCE, AND TOO OFTEN PATRON INDIFFERENCE) has fueled this runaway train.

**This is however intended to be a SPIRITUALLY ANCHORED HANDBOOK,** appealing to thought and reason, rather that emotion and feeling. Building from KNOWLEDGE – information, through EXPERIENCE. Adding trial and error, and practice, practice, practice, to WISDOM. Promoting performance maturity, a workman that need not be ashamed, RIGHTLY DIVIDING (APPLYING) THE WORD OF TRUTH.

*"OTHER* FOUNDATIONS *CAN NO MAN (OR WOMAN) LAY"*.

**First and foremost**, let me tell you a bit about yourself.

The greatest creative gift that humanity has been given is the **"air we breathe"**.

Fact #1

We have a factual record of creation in the first two chapters of the Book of Genesis in the Bible, the inspired Word of God.

Fact #2

We have many thousands of interpretations of the Bible available in our world today, but they all claim to be anchored in the Pentateuch, the Torah, the written Law of Moses. The many variations in interpretation occur much later in the record of scripture.

**Fact #3**

Each of us is intitled to our opinion! But, none of us are entitled to a personal interpretation of facts.

Fact #4

As God completed each of the first 5 days of creation, He surveyed all He had accomplished and declared, "It is good". On the 6th day God created man, and as he looked back on his accomplishments, he declared, "It is very good".

Fact #5

According to the record of the creation of man, Genesis 2:7 from the Hebrew Inter-linear Bible shares (the H numbers following the words are Strong's concordance numbers that give us the Hebrew word, and its meaning),

> **Gen 2:7** And the LORD$^{H3068}$ God$^{H430}$ formed$^{H3335}$ (H853) man$^{H120}$ *of* the dust$^{H6083}$ of$^{H4480}$ the ground, $^{H127}$ and breathed$^{H5301}$ into his nostrils$^{H639}$ the breath$^{H5397}$ of life;$^{H2416}$ and **man$^{H120}$ became$^{H1961}$ a living$^{H2416}$ soul. $^{H5315}$**

The word that is critical to this narrative is the last word in this verse

**soul. $^{H5315}$** and the Hebrew word and its meaning.

## REMEMBER

This book is not being written to **those who are just seeking a Job.** This book is being written to share with **those who want a Job *with* Job Training, so they can "earn while they learn".**

But employers are currently rarely known as training institutions. They may support training and reimburse for successful completion of blocks of study, but they are most appreciative of an employee's personal initiative, and third-party delivery resources.

To get off on the right foot we want to be up front and share without fear of contradiction that every TRAINING INITIATIVE regarding every CATEGORY OF WORK begins with a unique but specific glossary of terms (words).

Words by which to talk the talk. And action verbs that enable the description of the tasks associated with walking the walk – doing the work).

Words used to set forth educational learning objectives are not words of work. Typically, very small variations are not observable, and they are not action verbs, so they are not measurable.

- **Knowledge** "involves the recall of specifics and universals, the recall of methods and processes, or the recall of a pattern, structure, or setting."

Accessing and studying the following will give you a perspective of the many words available, but careful and thoughtful study of these will give you a sensitivity with regard to what words evoke physical activity and action. Randomly select a dozen words you feel imply action. Then write out in fifteen to twenty words what action each word would inspire. Read on to the next exercise and make comparisons.

https://www.learnersdictionary.com/3000-words/alpha/a/1

- **Comprehension** "refers to a type of understanding or apprehension such that the individual knows what is being communicated and can make use of the material or idea being communicated without necessarily relating it to other material or seeing its fullest implications."

Once more, access and study the following action verb glossary. If you did a reasonable study of the wordlist under knowledge, and you listed a few words as instructed, look up the same words here, and compare the use explanation here with what you came up with on your own.

https://mgt.buffalo.edu/content/dam/mgt/Career-Resources/Documents/ActionVerbs.pdf

Take a deep breath now and meditate on what you have been introduced to thus far. There's more to come, but unlike education, training doesn't have to watch the clock, or even play out on a schedule.

The purpose of training is performance, the successful mastery of each step, no matter how often the procedure or task must be addressed. But the true key to training success is the patient demeanor of the specialists that are evaluating the performance of a practitioner. There is a dramatic difference in the behavior of someone that is nervous, and someone that is unprepared.

## END OF THE PREVIEW OF NOTHING WORKS

THE BIBLE SHARES, "Other foundations can no man lay than that which is laid in Jesus Christ." No matter how we may try to explain it otherwise, Training is truly learning by example, and as trainers we must lead by example.

The purpose of training is to improve the life and lifestyle not only of the performer, but virtually every person in the circle of influence of the person being trained. The home, the church, the neighborhood, the employer's business, and ultimately the economy being served by business in general.

So, from this beginning, we began to build.

## BOOK 1 OF A 7-BOOK SERIES

## POSSIBLY YOUR BIGGEST DIY PROJECT YET

### *A CHALLENGE TO EARN WHILE YOU LEARN SKILL ACQISATION INITIATIVE*

PREFACE

We are living in times, and under conditions and circumstances in America today that have never been experienced before. Not in war years, the great depression, or during of the economic and social upheavals from the 50's forward.

In two recent books we have shared two sides of our current story. In one book, we shared **"Prison Bible College Stories"** and focused on the need for reform when lives and lifestyles run contrary to societal acceptability.

Unfortunately, reform does not come from isolation and banishment. The only true hope for the incarcerated is spiritual transformation, and that is the goal of initiatives like the **Prison Bible College**.

There are many volunteer ministries that dedicate and devote their energy and talent to offering "worship services" in jails and prisons across this great land of ours, but when it comes to the discipleship that is essential to transformation, it falls as short as throwing money at missions.

The intent of this dialog is not to discredit the efforts of these faithful volunteers, rather to share that the incarcerated need participation, rather than mere preaching. They need a challenge rather than a lecture. They need a continuum, rather than a mere glimpse.

The Bible shares with parents that they should *train up a child* in the way he or she should go. That doesn't mean merely how to get along at home. Nor does it mean how to get along while in school. That proverb means train them up with life skills (**how to live**) yes, but it also means "**how to make a living**". Workplace ethics, manners, grooming, appropriate dress, and of course basic performance skills.

Too many of the 2.3 million incarcerated in America testify that they didn't have that training. They came from broken homes (single parent), were raised by grandparents, grew up in foster care, or simply lived in gangs. Many of these that were not trained up may have acquired basic survival skills, but they almost all lack social skills.

The focus of the first book then was on *transformation (a change of heart)* brought about by *a change of mind*. Another proverb declares (paraphrased) as a person thinks, so are they. **Training up those that had been overlooked**.

The second book **"Nothing Works if you don't"** targets a totally different population, with totally different problems and circumstances. This book focuses on those millions that are *unemployed, or worse, under-employed*. Unemployment is a circumstance. Under-employment is an indefinite sentence, like parole.

There is some relief offered to the unemployed by means of social programs like unemployment benefits, food banks, etc. There is little or no hope for the under-employed. They are typically capable of so much more than their current employment requires, but they are locked into a wage that doesn't meet even basic expenses; their work environment and schedule don't permit job search activities, and their demeanor and attitude often reflect hopelessness.

In the book "Nothing Works if You Don't" a substantive review of America's historic skill transfer methodologies is offered; so, there is no value to be gained in repetition here. But the evidence based, time proven **"pay-for-skill"/" earn while you learn"** procedures offer the most immediate mutually profitable initiatives for the employee and employer.

The national problem is a lack of knowledge regarding the availability and economics of "Related Training" that is focused on workplace skill transfer today. There are very few social advising agencies that recommend this type training, because of their governmental association with traditional educational facilities.

Scriptures challenge, "Come, let us reason together" (paraphrased). So, it is to this end that it becomes essential to review the issues that keep educational methodologies from delivering relevant and timely workplace skill transfer.

1. Federal and State Educational Guidelines are based on approved curriculum and professionally credentialled faculty, to be eligible for funding.
2. Scholastic testing is criterion referenced and nationally normed.
3. Local fiscal accountability mandates economic class size (typically 12 to 15 enrollments per offer).
4. Trade related programs provide laboratory activities but on simulators, with quick connect and clock hour achievability to enable schedule compliance.
5. Few professionally credentialled faculty members have been equally qualified as task, tool, and trade performance proficiency.

Why are these problems? Because education typically addresses.

1. What (the task in general terms)?
2. Who (the person - typically a generic trades-title like electrician, mechanic, carpenter, plumber, etc.)?
3. When (typically a reactive response – like during a breakdown, in the event of an accident, following a disaster, or such)?
4. How much (this is usually discussed as tuition, fees, travel expenses, schedule inconveniences, even to the point of requiring a job changes or missing semesters because of scheduling requisites)? ... **Educational debt.**

What is not considered to be an educational responsibility?

1. How – (Tools - selection, use and care, and Techniques - the proper use in accordance with time proven best practices)?
2. How much or How little - (Techniques – compliance with manufacturing specifications, conformance with Regulatory and Safety mandates)?

3. Why – (Functional reason for intervention, motivated by local management strategy, or inventory practice)?

4. Why now – (during downtime/either reactive repairing, or pro-active preventive; or during runtime/monitoring and inspecting, servicing, or cleaning to avoid undesirable interruptions)?

5. Who – (to turn to as a mentor, coach, or consult when the task skills are beyond current training)?

6. What about Debt? There should be none. What is being shared here is that in a structured **pay-for-skill** initiative, the student is **earning while they are learning**, and their wage level is typically at a level from two to two-and-a-half times the area minimum wage. (don't expect this in an area that has opted to go to the fifteen dollar an hour minimum.

And as you might expect, and the list goes on. Without pointing fingers, it should be obvious that the difference is the need for real-world, real-time, real-task, real-tool, real-technique training on real-equipment, that is running real-processes, making real parts.

None of these real aspects of work will ever be accomplished to the level of national skill standard expectations in a laboratory, in a one, two, or three hour sessions, when the pay-for-skill participant has an eight hour straight time day, and if it isn't done right, the task can be reassigned for tomorrow, and as many tomorrow's as it takes to get it right.

SO, what is required for a structured Pay-For-Skill program? Very simply, two components distinguish such a program.

1. A related instructional source – (a means of attaining relevant knowledge [need to know], in a sequence that enables upward mobility, focused on the activities of the work discipline being studied).

2. A performance evaluation process – (involving a full spectrum of performance objectives as set forth in the national skill standard database; and proven performers that are accessible to

observe and evaluate the performance of each student against the national skill objectives).

Any employer should be open to hiring a self-motivated candidate for employment that has done "due diligence" and prepared themselves to be a worthy candidate for placement in a pay-for-skill program targeting one or more of the company's most crucial skill categories. With all of this having been said as a preface to this book, let's set sail and find just how enjoyable the journey might be.

The rest of this book will unlock the simplicity of sourcing both of these ingredients, and unfortunately, that only leaves one problem to be dealt with. The candidate for training. It isn't the absence of skill as much as desire, will-power, perseverance, and motivation.

## Challenge

There has never been a better time to take advantage of a proven process.

There has never been a more convenient and economical means of accomplishing what is perceived to be the most difficult and undesirable pursuits. Resources and tools have never been more reliable.

There has never been a greater need for semi-skilled and skilled talent.

## Questions

The question, "Can a willing employer have confidence in investing in you?"

Another question, "Are you willing to invest in preparing yourself with valid requisite credentials?"

Still another question, "Do you have the patience to serve two, three or four years to gain crafts credentials?"

Another question, "Are you comfortable coming home with dirty hands rather than a headache?"

**BOOK 2** OF A 7-BOOK SERIES

**I'm Glad You Asked About ...**
**THE BODY (THE MACHINE DESIGN)**
**WE'LL GET TO YOUR QUESTION,**
**BUT FIRST THINGS FIRST**

## PREFACE

In an introduction to this series of self-help Do-It-Yourself books, we addressed the plight of the unemployed, and to a greater extent, the under-employed, in a booming economy.

Unfortunately, as we entered February 2020 that booming economy plummeted when America fell victim to the corona virus pandemic. But as trite as it may seem, this too shall pass.

Circumstances beyond our control need not be a disaster, they can just as easily be an opportunity. We may not be at liberty to go to the streets, or shop at will, but we can use our time to share, or engage in self-evaluation, self-improvement, or personal development activities.

For my part, I am writing this self-help series, so that you can engage in a low-cost initiative to assure you or your family, friends, or any in your circle of influence that might be suffering from this temporary inconvenience, can assume accountability for personal development.

Never in recent history have we heard of the number of unfilled jobs in America exceeding the number of job seekers by nearly one and a half job seekers. The unfortunate truth is that available job seekers do not currently possess the performance skills required by the unfilled jobs.

Not only is that being a matter of the evening news now, it has been a continuously spiraling trend since the early 1990's when Industry began to seriously close down their trade schools, in favor of supporting the community colleges.

We closed our introductory work by sharing that everyone is entitled their own opinions, but no one is entitled to their own facts.

With that thought in mind, let me share that Industrial Reliability Services (maintenance, tool room, stock room and gage lab) terminology, tasks, tools, techniques and tactics are not concepts common to academia (education). That is fact, not opinion.

These concepts are anchored to national skill standards for personal performance, and are basically articulated across all fifty states and most of the developed countries of this world.

Educational standards on the other hand may address national and state learning objects. But the development of curriculum, selection of textbooks and supportive instructional support materials are all selected by local curriculum committees. Then the delivery is totally skewed and determined by the faculty teaching the course.

In large systems when multiple sessions of the same course are taught by different instructors, there will be significant variations in the student comprehension. Possibly the most influential impact on the consistency of the outcomes is the practice of grading student performances on the bell curve for whatever reason.

That too is fact, not opinion. But there is one additional distinction between training in conformity to the national standards, and the traditional scholastic schedule. Training is not dependent on the clock and class hour constraints. Nor is training a one-time lecture, test, or lab experience. In a performance-based skill training initiative, "If at first you don't succeed, try, try again."

Skills are seldom taught, they are caught (repetitively experienced and practiced until mastered). That expresses the fact that we don't master a skill by reading about it. We are typically introduced to such by reading, listening, and viewing supplementary pictures, prints, or even videos. But as the Eunuch of Ethiopia responded to Philip the Deacon (in the book of Acts in the Bible), "How can I understand what I am reading, unless someone explain it to me."

# NOW TO **THE BODY** (THE DESIGN)

## Footings

This is a field of endeavor that many of our society knows nothing about, because although footings are often constructed in open daylight, and not shielded from sight, they are often deep into ground, or down to bedrock, the area is dirty, and as a construction site, usually barricaded to keep wanderers from exposing themselves to the inherent dangers.

Focusing on the goals of this book, let's start with the dictionary definition of a footing. Dictionary.com shares … Footing (Building Trades – Civil Engineering) … the part of a foundation bearing directly upon the earth. A more general definition shares that a footing … the basis on which something is established or operates.

These definitions as simple as they are, illustrate the concept that we are discussing. The general definition might be compared as "politically correct", it has not real specificity, and it certainly doesn't offend.

The civil engineering definition however, says it is the part of the foundation that is in contact with the earth. It is left to the reader to assume that it is in contact with the earth in such a manner as to be load bearing. If possible, the most desirable footing would rest securely on bedrock. But in any circumstance, the footing must be capable of supporting any planned superstructure.

Typically, once a footing is in place, a foundation is added, and then the dirt is backfilled against the foundation so that the outside is never seen again unless it is dug up. The inside of the foundation is commonly known as a basement wall.

Going back to our human analogy, the foot, the load bearing member, the lowest part of the structure, the part that is against earth (tera firma), is the footing. The leg then is the foundation.

As we have all experienced at one time or another, the leg is absolutely helpless unless the foot is on solid ground. Try to walk comfortably in loose sand, or through several inches of mud or water.

There is one obvious difference between the human anatomy and the machine. The human anatomy is capable of moving as an assembly.

The footing (foot) foundation (legs) and frame (body) of a machine are intended to be stationary, and in most instances must stay stationary, or the process integrity will be compromised.

That brings us to another crucial consideration with regard to our analogy. **The frame of the machine must by anchored** to the foundation in such a manner as to withstand any and all process stresses. The process and products involved in anchoring the frame take us into a **need to study fasteners.**

1. **Forces** – reactive forces in that the frame and foundation have to resist any process forces so that work can be accomplished on the product. If the frame or foundation moved, the intended work would be compromised.

2. **Torques** – twisting forces, again reactive, and twists because not all work is done in a linear manner. Hold your arm out with a glass in it. You can twist your wrist clockwise or counter-clockwise; or you can raise your wrist up or down. Then too you can involve your elbow, or your shoulder. Every motion produces different twisting forces at the body level that must be resisted. If there is no resistance, there would be no relative movement, and without movement, there is no work.

3. **Mechanical shocks** – there are two distinct types of shocks that machinery can encounter. **Internal shocks**, produced by the process (such as when a stamping press first hits the product and goes through its forming stroke), or a load being transferred hits an end stop. And, there are **external shocks** most frequently caused by an accident (a fork truck or material handler misjudges the position of a load and runs into the machine), or natural such as an earthquake.

4. and **Vibrations** – generally the concern is for process generated vibrations. And, any moving part has the potential of generating vibrations. A motor with its coupling and the connected component must be accurately balanced, or a harmonic vibration will be continuously generated. A cutting tool advancing into a

workpiece at an improper feed rate will produce a vibration of a totally different frequency.

There are terms used in these descriptions that are not common street terms, but to encourage you, that is the reason for the books that follow, and that is also one justification for a formal pay-for-skill training program that provides enough time, and permits enough practice to master these terms and concepts.

# BOOK 3 OF A 7-BOOK SERIES

## I'm Glad You Asked About …
## Industrial Power Transmission
## You Can't Stand Still and Progress

## PREFACE

The singular purpose of Industrial Power Transmission is to perform work so that something of value can be done. But one of the facts of our physical world is that if there is no movement (motion) there is no work.

This is the third book in this series subtitled "I'm glad you asked about _____", enquiring about differing concerns that must be dealt with in relationship to craft and trade concepts.

Concerns that must be integrated and balanced to provide power and control of automation. Thus, enabling the economic production of domestic and military goods and services.

These books are not intended to be novels. They are in fact intended to be recruiting tools, that are focused on encouraging anyone. Specifically, those in our society that are unemployed, or under-employed; encouraging them to consider a self-directed career skill acquisition and enhancement pursuit.

It is the intent of these books to tell anyone that is interested, "What is *available, and* what it is *possible to do*". It is not the intent of these books to tell anyone "*what to do*".

It is the intent of these books to share concept by concept the value to be added by self-motivated individuals regardless of race, creed, color, ethnicity, or religious bias, becoming involved in the acquisition and upgrade of performance skills.

Hard and soft skills that are so essential to manufacturing reliability support and servicing. Semi-skilled and skilled positions more commonly identified as facility maintenance, machinery maintenance, and tool maintenance skills; inspection and quality control, etc.

It is also the intent of these books to introduce and heighten the awareness of the value of *pay-for-skill* or other forms of training identified simply as *earn while you learn* initiatives. The advantages of these forms of training are results based,

- You do not incur educational debt (college debt)
- You are not studying to pass a test
- You are training to perform tasks
- You hire into this type program at a rate nearly two times the normal entry level wage level
- You have insurance and benefits all the years that a college student is studying, preparing to hire into a position.

We have gone through a period of more than a quarter of a century with a national expectation. That the community colleges on a local basis were going to produce the skills needed to sustain industry needs in the geography served by each college.

This has not proven to be true for many reasons that are not relevant to this discussion. The nearly total absence affordable skill training in nearly every community or America is, however, our motivation for writing. We have been part or the silent majority far too long.

## YOU CAN'T STAND STILL ...

In our socio-economic culture, unemployment, and especially under-employment is similar to an addiction. You suffer through a setback, and then another, and soon, it seems addictive.

There are so many people that have a job, but their pay rate is inadequate for them to even think about any form of self-help, or educational pursuits.

The truth is there are a large number of self-help programs available on the internet for anyone willing to commit to the time and attentiveness to begin the skill acquisition venture. If you don't possess a computer, access can be gained through the use of computers in a local library. And, if you need assistance using the computer, the staff of the library is skilled and will assist you in a meaningful search.

The most significant needs, for any position of meaning in any industry today, are a demonstrated competency in math up to and including trigonometry; and an understanding of technical physics, focusing on electrical and mechanical basics.

Don't wait for a program, and above all, don't wait for someone to make the decision for you, if you are sick and tired of being sick and tired of repetitive layoffs (unemployment), or never having enough of a paycheck to meet basic living needs, make a commitment to invest time in yourself, and join us for a journey.

## YOU CAN'T STAND STILL, AND GET AHEAD.

In a previous book (Book 2) we worked through a math exercise that began with the number line, the place value of numbers, and then a review of how we apply numbers in our daily activities. We didn't just work problem after problem. We applied the numbers in context, so that we could communicate with clarity.

We looked at the many hundreds of numbers and sets of numbers we use in our daily activities and learned that will all those differing uses all we have to work with is 0, 1, 2, 3, 4, 5, 6, 7, 8, and 9.

We also learned that numbers themselves cannot talk. They merely give value to what we are talking about. In essence, in that discussion we even considered multiplication and division buying bolts or screws and then using them in quantities of six or twelve in an assembly operation.

So many people shy away from any involvement with factory work, and specifically with maintenance or any other term that sounds like maintenance. They don't want to get dirty. But honestly, dirt washes off, and companies that have dirty jobs pay well for those that will make the best of it.

In this day of robo-calls and offers of everything from insurance to car warranties, to every kind of get rich quick schemes, we must all be wary of wrong motives. Too many simply want all of your personal information.

We are not going to tell you that a self-directed career preparation initiative will be easy, but we can tell you we will take the mystery out of sourcing the means of accomplishing your goals.

It will not be easy, but remember what you are going through at the present time isn't easy either. You tolerate it, because you don't feel you have many options. Think about this, anything "*worthwhile*" is worth working for; and we submit that "*the worthwhile that we intend to elevate is you*".

In an earlier book, we shared that our role in providing this advising series is similar to that of a coach. We cannot play the game. We can merely teach, and challenge, or coach. You have to pay attention, and then follow the game plan as it has been shared, and then play to the best of your ability. Remember too, sometimes you win, sometimes you build character.

As we turn our attention in this book to Industrial Power Transmission, we are going to get to the nitty-gritty of motion physics, the science of movement, and the problem of friction. We have discussed a number of the concepts of this exciting field of involvement in earlier books of this series. Now we are going to get down to where the rubber meets the road, as they say.

One thing we will assure you as we begin this portion of our journey, if you climb on and stick with it, you cannot fail. You will rarely see a qualified performer out of work. Even during such a disaster as the coronavirus pandemic, with manufacturing shut down and plants closed, skilled performers find temporary work doing handyman projects in the area around them, or even signing on with a contractor assisting in storm cleanup, or taking a well-deserved rest (they can afford it for a few weeks).

I cannot tell you that you can be successful in the field of manufacturing if you prove yourself proficient in performing the tasks of any of dozens of job titles associated with production reliability. You will have to find other coaching if your desire is to play in a different game. Ok, pep rally is over, let's play to win!

## BOOK 4 OF A 7-BOOK SERIES

### Hydraulics, Pneumatics, Lubrication, and Water

## PREFACE

### What do you think caused that?

I was acquainted with an optometrist many years ago, long before the insurance mandate for patients to see a primary care physician and be referred to specialists. This optometrist declared emphatically that he could examine the eyes, and evaluate the total health of a patient. He added to his declaration that his diagnosis would be more accurate than any other medical procedure available at that time.

Over the years of our association, we witnessed his repeated recommendation of his patients to see a cardiologist, or a specialist in hypertension, a dietician, or a urologist. That always seemed to border on what we now call scams.

But believability grew over time, because his recommendations were always further diagnosed as a need for treatment in that area of specialization. And, believability was enhanced when we found out that our friend never offered the name of a specialist, and when the specialist asked the patients who referred them, they did not have any clue who our friend was.

I always thought that amazing. Then I went to a noted orthopedic surgeon for an evaluation for a potential knee replacement. After a few minutes getting acquainted, the doctor examined my knees, and then asked me to take off my shoes and socks. That seemed a bit unusual, but I did so, and he spent a good 5 minutes: examining each toe nail, between the toes, the bottom of the foot, and then back to the toe nails.

He then excused himself for a moment and when he came back, he opened a small leather case, took some tools out and proceeded to give me the most professional manicure anyone could ever hope to get.

When he finished, he sat back in his chair and said in a soft but firm voice, I can help you with the toe fungus, but I will decline the opportunity to operate, I believe you would have difficulty healing up, and would need routine physical and balance therapy.

Then the doctor said, you have a circulatory problem, and that is causing the fungus growth, but it is also the reason I feel inclined to advise against surgery. So, to make your visit worthwhile today, I will give you this advice; keep Vicks on your toenails, and the fungus will ultimately go away. He was right, but it was certainly not the outcome we expected.

Recently, another surprise. I visited an Eye, Ear, and Nose and Throat doctor to investigate the potential of inner ear issues causing balance instability. He investigated the ears from a visual perspective, and then inspected the nasal cavities and sat back and said, you've been Plavix for some time now, haven't you. Note: He did not say blood thinners. He said Plavix. Once more, I was amazed that a trained professional could render that specific a conclusion.

## WHAT IS THE POINT?

You have undoubtedly heard the expression, "Jack of all trades, but master of none"? That is so typical of a diagnostic technician, extremely capable of monitoring and predicting potential problems. But incapable of demonstrating mastery in any of the performance categories that he prepares work orders for.

This is book four of a series that focuses on the differing categories of the activities or work associated with manufacturing automation as we know it in the year 2020. We have long since moved away from an on-site repair shop. On one side of the shop, a mechanic was repairing a hydraulic valve, or cylinder, either to go back into inventory as a backup for those identical models that were in production.

In a corner of the shop, an electrician was rewinding an electric motor that had failed due to overload. On a bench near the motor rewinding station, another electrician was training a student in the procedures and tools used to replace contacts and restore electrical relays to their factory status.

And then in separate room off to one side, we would find another mechanic testing a hydraulic pump that he had just rebuilt. He had his own room because the shop complained about the noise he was making.

Test stands do whine and moan. But the audible distinction "between normal and oh-oh sounds" is one of the acute diagnostic skills that the technician must learn if he intends to be capable of intervening before products reach failure mode.

We won't find much inhouse repair any more, there are trade dedicated repair companies in every geographic area today, so industry reduces their need for manpower by sending materials out for repair or replacement. The cost of repairs is usually 65% of the cost of replacement product.

Ironically, many of the repair shops also serve as distributors for new product as well. So, it is often easier to replace rather than repair. This is especially true in the field of *fluid power, lubrication and liquid handling* products.

In the era of the inhouse repair, you would seldom see a repair and rebuild mechanic out on the production floor. Likewise, it was just as uncommon to see a shop floor mechanic meddling in the repair shop. First and foremost, their talents (skills) had been perfected differently, with a differing objective in mind. And secondly, the task activities require a differing inventory of common tools.

Today, in the year 2020 AD, the need is for a trained ***diagnostic mechanic, with a monitoring mindset. Especially in the fields relating to hydraulics, pneumatics, and liquid management** products and systems.* One reason, it has seldom, if ever been done this way before. In the traditional manner, if someone needed training in hydraulics, the maintenance manager would use some of his training budget and send the candidate to a vendor school somewhere. Guess what that person learned. That vendor's products, whether the candidate's company used any of them or not.

Such classes are taught generically. The instructor (usually a company field service representative) says, "This is the way we have always done it". To add to that problem, the demonstrations in class are often new product, and the instructor has neither the tools, nor the inclination to take the product apart and do surgery on the intricacies. If questioned, he or she might find a replacement part, but that is only half of the story.

It is like hearing only one side of a telephone conversation. Or finding the very last piece of a picture puzzle. You know it is the last piece because you only have one piece, and one hole. But you still have to study it to determine how it fits in. Seeing the one-part leaves everything else up to imagination.

So, the point, and the focus of this book, is to encourage *a different mindset*. A mindset that says, "I will learn as much as I can, and do as much as I am capable of doing, reliably, with what I have learned."

When I do not know, I will not be embarrassed and remain silent with guilt or shame. I will acknowledge that *I am still a work in progress*, and seek the assistance of someone that has been trained beyond my present level.

Then, instead of returning to the days of doing it all inhouse, being *a jack of all trade tasks*, I will be an advocate of sending repairs out, but *monitoring* the repair/replace decisions, to conserve on purchased services budgets.

With all of this having been said, let's conclude that a specialist is, "a jack of one trade, not trying to master them all", but for the foreseeable future, the opportunities for employment and pay-for-skill (earn while you learn) initiatives trend toward technician competence in cross-skilled and multi-craft disciplines. The big three in industrial manufacturing are electrical, mechanical, and fluids.

This is the first of three books in this series, and we are starting with fluids first, because the output shaft of the actuator, or motor, is the place that we have to monitor movement, and determine output horsepower. This is exactly backwards from tradition.

The common approach is read the nameplate on the pump motor, and that tells you *what this system is doing*. No: Wrong! That just tells you **what the system may be capable of doing**.

Once we deal with the muscle, we will deal with the mechanization that makes it all happen in our next book, and then we go to the beginning and consider where and how we get the energy to do all this work (our third book with trade focus). In that book, we will discuss Electricity with specific emphasis on the application of motors, electro-mechanical motor control (old school), drive technologies, and the pilot and control devices essential to automation.

**BOOK 5** OF A 7-BOOK SERIES

**SURE, THE ROBOTS ARE COMING ...
BUT WHO IS GOING TO MAINTAIN ...?
ARMS, LEGS, AND
ARTICULATION**

## PREFACE

In the beginning ... I keep coming back to that phrase, over and over, because without knowledge of where something began, and what its intent and purpose was, we have little or no clue what we truly have now.

A famous scientist once declared that *if you can measure something, you know something about it.* Some things cannot be measured with a ruler, or a scale though, so we have to find other means, and substitute the knowledge and data acquired thereby, for physical measurement.

One such alternative means of measurement, is TIME. Although time doesn't give us a specific numeric output like a ruler, by means of observation and data comparison, we can track the "timed-rate-of-change-of-_____".

- Condition
- Position
- Shape
- Size
- Volume, Etc.

## TEN YEARS LATER

Typically, there has been enough change over a ten-year period that basic modifications have been made to the machinery, the automation, the tooling, and even the product. Rarely are any of these changes, trivial or significant.

Documented thoroughly, or accurately enough that anyone unfamiliar with the actual activity could discern just what was done, and why.

Based on leadership, performance, or influence, there are other substitutions that can be made for physical measurement with regard to automation, industrial machinery, equipment, process/recipe, and even personnel.

Before I tell you the key to non-physical measurement of variability, let me ask you, "How far back (generations) can you trace your ancestry (parents, grandparents, great-grandparents, great-great- etc.)?

Did you know that the born and bred, full-blooded Jew, can typically trace their ancestry back nearly 3500 years to the great-grandsons of Abraham? Ask them, and they will tell you, "I am descended from the tribe of _____." (One of the twelve sons of Jacob, Abraham's grand-son).

Or, would it surprise you if I told you that *4 disciples of Jesus and 1 close friend wrote 4 books and one chapter of the New Testament* in the Bible. Their purpose was to review the message (history) of the entire Old Testament. And, to relate the events of the beginning of the first century AD that gave birth to "The Church", the universal body of believers. *But, a first century Jewish Historian (Josephus) a contemporary of the disciples, personally wrote 20 volumes, to share the secular perspective of the history of the Old Testament, and the birth of the church.*

Does that give you, any clue as to the key to the measurement of non-physical entities? There are literally many procedures, but the key to the success of any, or even all of them, is ***DOCUMENTATION***!!!

I can't emphasize that enough, because very little or none of it is being done as a result of corporate, or organizational procedure. To substantiate my declaration of the importance of documentation, let me share that the average person will immediately puff up, grab a log book, and say, "I don't know about the rest of the world, but I keep good records".

Yet, if I ask that same respondent to show me the *maintenance history*, or the *purchased parts history*, or the *run time history*, or the *physical modifications and upgrades* made from the time of purchase to yesterday, for any given machine or piece of equipment on his shop floor, he would validate my claim.

If he still claimed to be current with everything, I would ask him about the *manufacturer's blueprints* showing how the machine or equipment was designed and built, and all the *change orders*, and *modification updates* with *dates and the names of the company personnel* that performed each activity associated with the change.

It is not hard to see that there are many language systems associated with the processes of manufacture, and each of them must be monitored, tracked and appropriately documented. The idea of paperless has permeated all of society, and more importantly, with our move to push manufacturing offshore to third world countries to boost economic leveling, much of the documentation we desperately need is not even in our own language.

That creates a great problem when we have to fly a service technician in from overseas to service a piece of equipment their company built, and when they get here, we find out that not only are we incapable of interpreting their documentation but we cannot communicate, because they do not speak our language.

Have you noticed a trend over the last 20 years, where traffic departments are putting up signage that has symbols on them instead of words? That is because of immigration, and tourism. A majority of newcomers and visitors do not speak our language.

I heard a discussion recently regarding the spread of the Word of God throughout the World, and the desperate need for interpreters, because there are more than 7000 languages in the World, and more than 4000 do not have a Bible in their own language yet.

The numbers are not as bad, but too many business leaders are under the impression that all communication is achieved either by oral or written communication, in person, by the airways, or through computerization and the press.

I submit that after a candidate for employment in the field of monitoring and managing manufacturing process reliability has accomplished *math*, and *technical physics* skills, that the next most important skill is *sketching with dimensioning, notes, dates, and his or her name.* There are too many things that are not obvious, and too many words that are not yet mastered to be capable of communicating in the industrial arena today without the ability to communicate in pictures.

If a person can't sketch, they had better become an expert in photography, both still, and video; and not just taking close up pictures, but using flash, and then getting their name and date on each picture. Too many, working in a manufacturing facility will attempt to describe a hydraulic pump as a black box, or a hydraulic reservoir as a storage chest.

Then there are those that see an electrical limit switch with a flat cover on it, and they think that someone put a step on the machine for them to stand on. Think about the conversation when someone tells a manager that someone bent the step on the side of the machine.

I think you get the picture. As we approach this, our fifth book, we are going to be talking about what goes on in automation. We have talked about footings, foundations, frames, fasteners, and actuators. Now we are going to talk about the automation that actuators push or pull, or twist around.

In the discussion of the way a hydraulic component and how a hydraulic system works, we went to the point of work, the output of the actuators, to determine what was expected of the system. We will not have that luxury in this book, because there many millions of things done by means of automation. What we will do, is discuss the very basics of automation, that will still apply in older, pre-1960 style automation.

I will share at this point, before we enter into this discussion, that there will be significant emphasis on the differing language systems, and the essential documentation needed to manage performance integrity in mechanical fabrications.

# BOOK 6 OF A 7-BOOK SERIES

## The Brain And Nervous System

## PREFACE

When we began this journey and offered our first book, "Nothing Works; If You Aren't Willing To!" we set forth two precepts. The first, that we are created beings, with the gift of life, and a purpose far beyond ourselves.

The second precept is that from an instructional and performance-based skill transfer perspective, we learn through experience. We learn best when we progress from what we know to continual revelation of the unknown, and progressive performance mastery.

I have not attempted to research the number of textbooks and internet offers concerning theory. We live with theoretical medicine, theoretical science, theoretical health and wellness, and theoretical financial security.

If I were to venture a guess, I would speculate that there may be 100,000 textbooks out there that deal with how to teach math alone. But textbooks are written by teachers, for teachers, and every author has a better way to teach, and every teacher adds their own experience and teaching style.

The problem is academic math does not meet the problem-solving needs of the blue-collar (craft and trade) worker on the shop floor. I graduated from General Motors, with an industrial engineering degree, and a mechanical engineering minor. I had academic math through calculus and differential equations.

But I share this without fear of contradiction if you follow the facts, my math education began my first day on the job as a maintenance supervisor in a foundry. I had 32 employees in my role as supervisor, with two to five skilled journey workers in each of seven trades.

I showed up with my slide rule and my book of engineering formulas, held my first departmental meeting, and after a few very harmless questions about why I came to work in the foundry, I felt well equipped to deal with the questions that this crew of trade mechanics would raise. That was my first official hour on the job.

In a matter of a month, I had been exposed to math that is not taught in the textbooks. Math that crosses over between the trades, but is so crucial to selecting replacement and repair products when there is no interchange available.

I was also introduced to tools that neither an Industrial engineer or a Mechanical engineer were never introduced to in the arena of formal education. Not just tools, but gages, instruments; test procedures, and shop floor process management.

## NOT AS TAUGHT IN TEXTBOOKS

Within a few months, I was introduced to foundry mold designs that were next to impossible to measure accurately, and when I asked how decisions were made regarding a change, I was told that a retired toolmaker came back on a contract, had a mold made with the pattern in question. He then dusted the drag mold with fine corn starch, closed the mold and reopened it. After studying it for about ten minutes, he took out a pocketknife scraped a vent in one spot, then told the tool room foreman to build the core up by .011 of an inch. The mold was then closed, and a casting was successfully poured.

The contractor did not stay long enough to see the core modified, and the next production run successful. Foundry management felt the contractor was arrogant in that he did not stay to see the results. The lesson I learned was the there is a dramatic difference between arrogance, and confidence. A famous baseball player once declared, "It ain't braggin', if you can do it".

Not only was this specialist confident, but when he sent his invoice it was less than half of what he had initially quoted. He stated that he found the problem, fixed it, and there was no need to stick around.

*That truly opened my eyes to the true meaning of skill, and performance-based outcomes.*

# ANOTHER, NOT IN TEXTBOOKS

About 18 months into my tour of duty in the foundry, we experienced an electrical storm in the area that knocked out power for a day and a half. When the power was restored, we began the processes of restarting operations. Within a few days, as we approached full operation, we began to trip current breakers on critical process motors (sand mixer motors, and cupola exhaust fume separator motors).

Many hours of testing were spent with the cooperation of the city utility, corporate engineering, and our own foundry staff. One of our production operators brought a recording chart from a sand tempering system and showed us that our control voltage level had been between nine and twelve volts below normal ever since we restored power.

With that data chart in hand, we consulted with the power utility and found that a major transformer bank had been damaged by a lightning strike during the storm. Checking further into the repairs made, it was found that an undersized-transformer had been installed, and the voltage level was proportionately low.

That was another eyeopener, "Ohms Law" ($E=I/R$) still worked, but the motors were drawing an excessive current, and the current overloads were doing what they were designed to do. And, without evidence to direct us otherwise, our maintenance electricians kept replacing the suspected current overloads.


# A CHALLENGE

Years later, as Program Chairman for Automated Manufacturing with a college in Indiana, I spent years trying to get our math faculty to accept the math taught in the electronics curriculum as equivalent math credit toward graduation.

## A CHANGE OF DIRECTION

Recognizing that my heart was in skill transfer, proven by performance, and anchored to the National Skill Standards for Related Instruction and Performance Objectives set forth by the US Department of Labor, I left the school. I joined my wife and we served as field representatives representing one of the leading suppliers of related instruction for apprenticeships. We actively worked with the sponsors of more than 1200 registered programs in the southeastern United States.

During this time, we were perceived to be in competition with the community college and technology centers. The truth is we were not competitors. We were an alternative supplier of choice for companies that had sought trade training. They were struggling with paying college tuition and not getting graduate with the job-task skills essential to their succession.

In late 1993, I suffered sudden cardiac death, had surgery for the removal of an abdominal tumor, and spent a few months recovering. Due to a spiraling downturn in industrially sponsored apprenticeships, the stakeholders in the company we worked with felt the need to do away outside field sales representation, and take the marketing and sales in house (telemarketing).

As soon as I was medically released to return to work, I joined a community college in the contract training, non-credit division, facilitating classes off campus, company by company. Due to enrollment and economic changes in the 1997-1998 time, my supervisor was directed to terminate all off-site classes, and bring the students to campus.

With no class responsibilities, I moved to a competitive college in another state, again in non-credit, contract training, and assumed responsibility for marketing and facilitating a new offering, Web-based online content available 24/7 via the internet.

We had great success for the first 9 years, but as we approached what is now discussed as the depression of 2008, the college system reorganized. Collapsing more than twenty-five campuses into 14 systems. Our campus administration was absorbed into another system, and my responsibilities were no longer needed.

Fully convinced that my attempt to get the significance of skill proficiency recognized by academia, I moved into contract consultative partnering, offering web-based training, and personally facilitating demonstration labs.

## MOTIVATION FOR WRITING THIS SERIES OF BOOKS

In 2018 we were blessed to be favored with an order to administer a four-year apprenticeship initiative for industrial maintenance associates and tool and die maker associates (32 apprentices in total).

Things went well until the coronavirus pandemic of 2020, and the plant shut down. The people are now scattered to the wind, and unless they make contact, absent from all services.

So, I am at home with 65 years of multi-craft and cross-skill training transfer experience. The only logical way to share would be to write about that experience. Now I have stuck my neck out for some new experience, publishing, and marketing.

Now that you know my story, let us get into the matter at hand. So far, we have talked about footings, foundations, frames, actuators, and automation, what more is there?

Purely and simply, the **brains** (*controls and controllers*) that decide what must be done, the **nervous system** (*pilot devices and other command signal devices and systems*) that says, "do it", and the rest of the **nervous system** (the field sensing devices – *switches and sensors*) that says, "it is done".

We addressed the nervous system briefly in our last book, as we introduced the ladder, the true backbone of symbolic control logic. For those who were not with us at that time, LADDER LOGIC is the most commonly used symbolic logic method of diagramming for electrical systems. The following websites offer a good introduction to ladder diagramming circuitry, and symbols charts.

https://www.plcacademy.com/**ladder-logic-examples/**
https://wiringforums.com/
**ladder-diagram-electrical-symbols-chart/**

**BOOK 7** OF A 7-BOOK SERIES

LIGHT AT THE END OF THE TUNNEL
**... *HOW TO DO*** WHAT NEEDS TO BE DONE

## PREFACE

**This is where the rubber meets the road.**

Through the discussions in the preceding 7 books, we have deliberated options, things that are available, things that are opportunities, and things that you can do, to enable you to pursue improvement of your status. All of these of course are dependent upon your willingness to act on the information being made available to you in this series of books.

My purpose in writing is to share the truth with you, because although much of what I share is widely known and acknowledged, there is no obvious effort to make it known with any desire to alleviate the plight of those who are not privileged to such, first hand.

My motivation is fueled by the fact that the market niche that I had devoted 40 years to was dealt a death blow by the promises of the US Department of Education and the Federal and State Legislative bodies. Promises to provide future workplace skill needs. To sweeten the offers, the government supported these promises by offering incentives to industry to close down proprietary trade schools.

**Do some math.** Since the late 1940's the U.S. Department of Labor's Bureau of Apprenticeship and Training catalogued and perfected the Related Training Instructional requirements, and the Job/Task performance skill requirements associated with 13,000 distinguishable and distinctive job titles. The purpose of this initiative was to provide a national resource for standardization in the processes of work, so that outcomes would be comparable when adhered to anywhere in the world.

The other side of the equation in this so-called math exercise is the fact that the Colleges and Universities nationwide offer half of that number of distinguishable certificates, diploma, or degree titles. Moreover, degree titles do not correlate with job titles.

Academic course descriptions say, "When you complete this course, you will be able to sort, list, catalog, and describe". And most academic learning objectives are based on memorization.

Apprenticeship objectives say, "When you prove performance mastery, you will be able to do …, and do it in accordance with quality standard …, and you will have accomplished this in a time range of …, etc. And all performance objectives are based on habituation, the outcome of repetitive practice and spaced reviews. *The only potential for failure in a structured apprenticeship is a failure to try (a character flaw).* All tasks are accomplishable, and all of the tasks the national database support are evidence-based and time-proven.

An article on apprenticeship was forwarded to me, and it has such a bearing on the focus of this discussion, as well as the timing. Please take the time to read it.

https://www.jff.org/what-we-do/**impact-stories/** center-for-apprenticeship-and-work-based-learning/ **apprenticeship-vital-manufacturings-post- pandemic-recovery/**

But there is also an article about the state of affairs with regard to education doing what was originally promised. I share it because I didn't write it but it supports this totally.

https://www.foreignaffairs.com/articles/ united-states/2013-04-03/**why-american -education-fails**

**Why American Education Fails**
And How Lessons from Abroad Could Improve It
By Jal Mehta May/June 2013

In his landmark 1973 book, *The Coming of Post-industrial Society*, the sociologist Daniel Bell heralded the United States' transition from a labor-intensive economy that produced goods to a knowledge-based one geared toward providing services. No longer could success be achieved through manual, assembly-line work; it would require advanced skills and creativity. At least since then, American politicians and (self-identified) subject matter experts have regularly stressed that education holds the key to the country's future. Everyone seems to agree that good schools are prerequisites for broad economic prosperity, individual social mobility, and a healthy civil society in which informed voters engage in the public issues of the day.

Although no one disputes the value of education, how the country should improve it is fiercely contested.

With that input, and national leadership we began to shun productive work to pursue information.

Think about our present situation, the virus pandemic, after months of shutdown and quarantine, many people do not want to return to work.

Consider point 6 and part of 7 from an article **"29 Ways the Educational System Is Failing Students"** posted on the website.

https://medium.com/@meholstein/29-ways-american-schools-fail-students-b0cf3fc805ba

6.  Schools do not hire subject matter experts to be teachers. *Instead, schools require a Bachelor of Education. A Bachelor of Education learns about teaching methods. They select specializations such as Elementary School, Middle School or High School.* Actual

expertise in their subject matter, if it's covered at all, is covered for only a few courses.

What this means is teachers are barely more qualified in their subjects than the students they teach. Any college graduate is more educated in their subject than the teacher of that subject at the local high school. This guarantees that teachers lack the depth of knowledge required to prepare students for future education. And teachers have hundreds of students a year.

7. Because schools only teach and assess the ability to memorize, students have no opportunity to exercise their intelligence outside of this. According to schools, the only way a student can be smart is if they can memorize well. Because this is how schools judge students, students learn to judge *each other* on this as well. They begin ranking their intelligence (and even their worth) by their ability to memorize.

I share this because I didn't say it, I read it, but it is consistent with the point I am attempting to convey in this challenge.

I suggested earlier that you **do some math**.

**Here is a serious question that relates to math**.

Why should anyone want to incur more and more debt pursuing more and more education, to acquire performance skills, when the schools are neither designed for or inclined to develop performance skill transfer initiatives?

I do hope by now I have your attention. If so, we will move on to what needs to be done, in what order, to get you from where you presently are, to where you desire to be.

Remember, when we started this series of books, I specifically shared that the information shared would be of extreme interest to the unemployed, and more specifically the under-employed. Let me elaborate on that statement a moment by sharing with you that if you find information in any of these books helpful in your particular circumstance, you may be one of the under-employed.

Be blessed,
Until we mee again,
Mel